PRACTICAL HYDRAULICS

PRACTICAL HYDRAULICS

ANDREW L. SIMON
Professor and Head
Civil Engineering Department
The University of Akron

CONSULTANTS

WILLIAM M. GLAZIER
Community and Technical College
The University of Akron
Akron, Ohio

ALBERT CHRISTENSEN
Mohawk Valley Community College
Utica, New York

John Wiley & Sons, Inc. New York · London · Sydney · Toronto

Library of Congress Cataloging in Publication Data:

Simon, Andrew L
    Practical hydraulics.

    Bibliography: p.
    Includes index.
    1. Hydraulics.   I.   Title.
TC160.S38   1976   627   75-28304
ISBN 0-471-79186-5

Printed in the United States of America

10 9 8 7 6 5

Book Design by Angie Lee

TO MARIKA, ANTHONY, and ANDREW

PREFACE

The science of hydraulics is old as civilization itself. The control of floods, irrigation, and water supply by ancient man required intuitive understanding of the basic physical principles of hydraulics. Yet, in spite of its age, hydraulics is still a somewhat uncertain science, full of theoretically indeterminate factors. In theoretical developments newer sciences have long eclipsed hydraulics, but its overwhelming importance in man's fight for survival is still unquestionable.

During the past century enormous progress has been made in the understanding of the fundamental laws of the mechanics of fluids. Powerful mathematical techniques are now available for putting these fundamental principles into practice. Yet, most practical hydraulics problems still defy these theoretical solutions. Practical hydraulics is perhaps as much an intuitive art as a science.

One of the reasons for this theoretical uncertainty of hydraulics is the large number of ill-defined variables that enter into even some of the simplest practical problems. The often unknown interdependence of these pertinent variables makes it impossible to develop reliable answers on the basis of fluid mechanics principles alone. Therefore to consider hydraulics as simply experimental fluid mechanics is a faulty oversimplification.

This book provides essential information for anyone seeking practical answers for everyday hydraulic problems. No attempt is made to present the material in the framework of theoretical fluid mechanics. The topics are presented as they appear in practice and not as examples of the application of certain theoretical concepts. Derivations are included only if their understanding is necessary for the proper use of their results, and they are kept at a level that does not require calculus.

The practical approach to the arrangement of topics resulted in the splitting of some subject areas that are conventionally presented side by side. For example, the concept of water pressure is discussed in Chapter 2, while manometers are presented in Chapter 10. Similitude is treated in Chapter 9 as a basis for model analysis of structures, while dimensional analysis is given in Chapter 10 as a way to reduce experimental data.

Some materials normally found only in graduate-level text books are given prominent treatment. For example, the chapter on open channels considers normal flow as a convenient simplification that may or may not be applicable to practical problems. The discussion introduces backwater and channel delivery computations, which rarely appear in undergraduate texts and never in those written without the use of calculus. The lack of analytic development and the detailed example problems will be appreciated by the practically minded hydraulician. Similarly, the chapter on seepage includes graphs that are generally shown in highly mathematical graduate texts. The lack of theoretical derivations is overshadowed by the practical usefulness of the solutions.

The chapter on fluid properties emphasizes the phase diagram of water. Other textbooks do not discuss the concepts of internal energy, specific heat, and latent heat. However, the practical aspects of ice formation and removal in rivers and hydraulic structures could not have been introduced in a meaningful manner without first discussing heat energy in detail.

The chapter on the laws of fluid mechanics offers an abbreviated treatment of these topics as compared to similar texts. The approach is to quickly introduce the fundamental principles and proceed, in succeeding chapters, to their practical applications. This tends to deemphasize the theoretical approach. Otherwise, without a judicious dose of hydraulic uncertainty, fluid mechanic principles lend themselves to endless theoretical refinements. With increasing theoretical complexity goes an impression of increasing precision and accuracy. Then, with the manipulative perplexities resolved, the student would have a false impression of understanding.

For the practically inclined student the chapters on pumps, hydrologic fundamentals, hydraulic structures, and flow measurements will be useful. The final chapter contains a survey of commercially available hydraulic measuring devices and gives their manufacturers. Although the currently developed devices may be quickly outdated, the only alternative was to end the discussion, as usual, after a consideration of manometers and pitot tubes.

Throughout, metric units are used alongside of conventional English notations. Many European design aids, graphs, and formulas were included, most of which have not yet been published in English. Converting them to English units did not seem feasible at a time when 90 percent of the world's population uses the metric system and when the United States seriously considers metrication. The use of metric terms in the text and examples will enable the student to learn the use of both systems at equal ease.

I thank Albert Christensen, William Glazier, Garold D. Oberlender, and Robert E. McGrath for their helpful advice during the preparation of the manuscript. I especially thank Pramath Pramoolkit for his diligent review of the text and for preparing most of the worked examples. Dorothy Guilliams carefully and patiently typed the manuscript.

I must express my love and appreciation to my wife, Marika. Without her constant interruptions, this book would be outdated by now.

Andrew L. Simon

CONTENTS

PRACTICAL HYDRAULICS

PHYSICAL PROPERTIES OF WATER

The way water behaves under various conditions encountered in practice depends primarily on its fundamental chemical and physical properties. These are actually controlled by water's molecular structure and by its internal energy. The terms and concepts introduced here will be frequently referred to in all later chapters.

INTERNAL ENERGY AND THE THREE PHASES OF WATER

Water is a chemical compound composed of oxygen and hydrogen. In each water molecule there is one oxygen atom for two hydrogen atoms. Clusters of water molecules are more or less bonded together by their hydrogen atoms. This type of atomic bond is called *hydrogen bond.* The degree of hydrogen bonding, the amount of energy holding the molecules together, depends on the temperature and the pressure present. Both temperature and pressure are manifestations of energy.

Energy is universally measured in joules. One joule of energy is the potential to do work of a force of one *newton* exerted over a distance of one *meter*. Heat energy is expressed in calories. One *calorie* is the energy required to heat one gram of water by one degree Celsius. (The English equivalent of calorie is the British Thermal Unit, abbreviated BTU. One BTU equals 1055 joules, or 252 calories.)

Depending on its *internal energy content* water appears in either liquid, solid, or gaseous forms. Snow and ice are solid forms of water; moisture, water vapor in air is gaseous form. The different forms of water are called its phases. Whether water is in its solid, liquid, or gaseous phase depends on the amount of energy held in its hydrogen bonds. In solid form all hydrogen atoms are bonded; in liquid form fewer hydrogens are bonded. There are no bonds in the gaseous phase. Water is a stable compound; the bonding between the hydrogen and oxygen atoms does not decompose until the temperature reaches thousands of degrees Celsius.

The amount of energy required to raise

1

the temperature of a substance by one degree is called *specific heat*. Specific heat of ice is 1.95, of water it is 4.19 joules per gram per degree Kelvin.* For water vapor the specific heat at constant pressure is 1.81, and at constant temperature it is 1.35 joules per gram per degree Kelvin.

This means that it takes less heat to warm up ice than to warm up water, and even less to warm up the same amount of water vapor.

The various phases of water as they depend on the energy content are shown in a phase diagram in Figure 1.1a. The two

coordinates of the phase diagram represent the two forms of energy, pressure and temperature. The diagram contains three zones within which water is in solid, liquid, or gaseous condition. To pass from one phase to another phase, either the pressure or the heat energy must change. Therefore energy is either added or taken away from water in order to change its phase. The amount of energy required to change from one phase to another is called *latent heat*. Figure 1.1b shows the concepts of specific heat and latent heat as they relate to the changes in tempera-

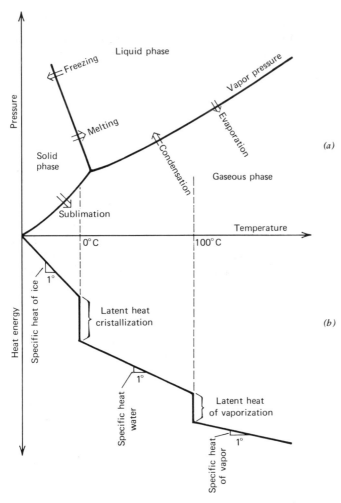

FIGURE 1.1 Phase diagram of water.

* The Kelvin temperature scale is equivalent to the Celsius scale plus 273.15 degrees Celsius.

ture of water held at constant pressure. To change from solid to gaseous form requires 2834 joules of energy per gram of water. This is called latent heat of sublimation of ice. In practice we see this process when snow disappears without melting during a cold but sunny, winter day. The energy input needed for this sublimation is supplied by the sun.

Melting of ice, to change from solid to liquid phase, is called fusion in science. Latent heat of fusion of water is 334 joules per gram. Freezing is the same process reversed. To freeze water, 334 joules of energy must be taken out of each gram of water.

To change the liquid phase of water into the gaseous phase we need a latent heat of vaporization of 2500 joules per gram. The process of evaporation is a complex one.

The heat content of a lake, the amount of heat energy entering through the inflowing waters, the velocity of wind carrying away moist air from the surface of the water, incoming solar radiation, and the temperature of the air all influence the rate of evaporation. At sea level exposed water surfaces are under the pressure of the atmosphere. The standard atmospheric pressure is 14.7 psi at 45° latitude and 68°F. If the temperature is raised to the boiling point (100°C or 212°F) water evaporates. At higher elevations the atmospheric pressure is less; hence water evaporates at temperatures lower than 100°C. The amount of pressure at which water changes phase from liquid to gaseous form is called *vapor pressure*. As shown by the line representing this boundary between gaseous and liquid forms in

EXAMPLE 1.1

How much heat energy is required to melt 500 grams of ice that is initially at −3°C temperature and raise the temperature of the water to +10°C?

SOLUTION:

The specific heat of ice is 1.95 joules per gram per °C. Hence to raise the temperature to zero degrees

$$1.95 \times 500 \times 3 = 2925 \text{ joules are required.}$$

Since the latent heat of fusion of water is 334 joules per gram, to melt the ice we need

$$334 \times 500 = 167,000 \text{ joules}$$

Since the specific heat of water is 4.19 joules per gram per °C, to warm the water we need

$$4.19 \times 500 \times 10 = 20,950 \text{ joules}$$

Therefore the total heat requirement of the operation is

$$2925 + 167,000 + 20,950 = 190,875 \text{ joules}$$

As 1 BTU is 1055 joules, this equals

$$180.92 \text{ BTUs}$$

One BTU equals 777.5 foot-pounds of energy; hence our energy requirement is

$$180.92 \times 777.5 = 140,665.3 \text{ ft-lb}$$

Table 1-1 Vapor Pressure of Water

| Temperature | Absolute Pressure[a] | |
°C	newton/m^2	psi
0	613	0.0889
10	1226	0.1779
20	2335	0.3384
40	7377	1.0695
60	19,924	2.8893
80	47,363	6.8676
100	101,376	14.6984
105	120,869	17.5239

[a] Standard atmospheric pressure is 14.7 psi absolute pressure, or 2116 lb/ft^2, or 32 ft of water. This equals 9.75 meters of water. In metric computations standard atmospheric pressure is usually taken as 10 meters of water.

Figure 1.1a the vapor pressure depends on the pressure and temperature. Table 1-1 shows this relationship.

In closed systems like pipes or pumps, water may change phase because of changes in pressure, even though the temperature remains constant.

In the English system of weights and measures the commonly used value of vapor pressure of water is 0.34 lb/in.^{2} at standard atmospheric pressure and 68°F. In everyday hydraulic design the vapor pressure of water is often of great importance. In the suction pipes of pumps as well as at the tip of the impellers the pressure of water often reduces to levels below vapor pressure. Water at those locations turn into vapor. As the vapor bubbles move on to points of high pressure, the bubbles collapse in a noisy and violent manner causing considerable damage to pump and pipe. This damaging process called *cavitation* is avoided by observing correct design procedures.

MASS, FORCE, AND DENSITY

The mass of water in a unit volume is called its *density*. Its magnitude is depen-dent on the number of water molecules that occupy the space of a unit volume. This, of course, is determined by the size of the molecules and by the structure by which they bond together. The latter, as we know already, depend on the temperature and pressure. Because of the peculiar molecular structure of water and the change of the molecular structure when the water takes solid form, it is one of the few substances that expands when it freezes. The expansion of freezing water when rigidly contained causes stresses in the container. These stresses are responsible for the weathering of rocks and could damage pipes or structures if their effect is not considered in the design.

Water reaches its maximum density near the freezing point at 3.98°C. Table 1-2 gives the density of water at different temperatures. In the English system density is expressed in slugs per cubic foot. One slug equals one pound force per one ft/sec^2 acceleration. The density of water in slugs per cubic foot at atmospheric pressure and at 69°F is 1.94 lb sec^2/ft^4.

As shown in Table 1-2 the density of ice is different from that of liquid water at the same temperature. This is why ice floats on water.

Since sea water contains salt, its density is greater than that of fresh water. The density of sea water is usually taken to be 1.99 slugs/ft^3, about 4% more than fresh water.

When one considers the concept of unit weights, considerable confusion may re-

Table 1-2 Density of Water

| Temperature | Density | |
°C	g/cm^3	slugs/ft^3
0 ice	0.917	1.779
0 water	0.9998	1.9406
3.98	1.0000	1.941
10	0.9997	1.940
25	0.9971	1.935
100	0.9584	1.860

sult from the difference of interpretations between the new S.I. units, the conventional English system of weights and measures, and some regionally used metric systems.

By Newton's law, force equals mass times acceleration,

$$F = m \cdot a \tag{1.1}$$

In the new S.I. system the unit of mass is the kilogram. Force is a secondary dimension defined by

force = kilogram
        × gravitational acceleration

and is called a newton.

In the English system we speak of *absolute or gravitational systems*. In engineering usage the gravitational system is predominant. In the English gravitational system the pound force is a primary unit along with foot for length and second for time. Hence mass is a secondary unit derived from Newton's law in the following form:

pound force = slug
        × gravitational acceleration
        = pound mass × (one ft/sec^2)

Some countries with metric traditions use the term kilogram to express force as a primary unit. In this interpretation density or mass is a secondary unit given as kg s^2/m^4. Although this practice is now discouraged by international standards, it is still widely used and can be found often in the technical literature.

In this book the new absolute S.I. system will be used alongside of the conventional English gravitational system. The resultant difference in interpretation of the *concept of weight* must be kept in mind. When we use metric terminology, weight of water is expressed as kilogram mass. Force, when specifically used, is given in newton (abbreviated N); 4.448 newtons equal one pound of force. This term will be especially important in connection with impact forces arising from changes of velocity of flowing water. Kilogram therefore will always refer to mass.

In the English notations pounds refer to pound force, a product of mass in slugs and gravitational acceleration.

The *gravitational acceleration* on the surface of the Earth averages between 9.78 and 9.82 m/s^2. For design the value of 9.81 m/s^2 will be used. In English units the value of 32.2 ft/sec^2 is used as standard.

For rough computations using the metric terminology the approximate value of 10 m/s^2 may be occasionally allowed. This introduces a small error, about 2%, but the ease of multiplying or dividing by 10 and the resulting elimination of possible computational mistakes offset the difference.

In the metric (S.I.) system therefore the value of the *unit weight of water* can be taken as 1.0 gram/cm^3. By virtue of the metric system, this value in ton (metric)/m^3 is also 1.0. Similarly, by definition, the unit weight of water is one kilogram per cubic decimeter, or one kilogram per liter.

In English units the unit weight of water is

$$\gamma = \rho \cdot g \tag{1.2}$$

where γ is in lb/ft^3, ρ is the density in slugs, and g is the gravitational acceleration, 32.2 ft/sec^2. The commonly used value for γ unit weight will be 62.4 lb/ft^3.

Practical work using metric units is decidedly easier than the use of the English system. Appendix I provides a short explanation of the proper use of the metric terminology.

EXAMPLE 1.2

A container weighs 3.22 lb force when empty. Filled with water at 60°F the mass of the container and its contents is 1.95 slugs. Find the weight of the water in the container and its volume in cubic feet.

SOLUTION:

One slug is defined as 1 lb of force per 1 ft/sec² acceleration. By virtue of Newton's law of motion

$$1 \text{ lb force} = \frac{1 \text{ lb mass} \times 32.2 \text{ ft/sec}^2}{g_c}$$

Where

$$g_c = \frac{32.2 \text{ lb mass-ft}}{\text{lb force-sec}^2}$$

  = a term serving the purpose of dimensional adjustment. From this it follows that one slug, by definition equals 32.2 lb mass. That is;

$$1 \text{ slug} = 1 \frac{\text{lb force-sec}^2}{\text{ft}} = 32.2 \text{ lb mass}$$

Then 1.95 slug = 1.95 × 32.2 = 62.79 lb mass. This under the gravitational acceleration equals 62.79 lb force. The weight of the container equals 3.22 lb force, so that the weight of water in lb force = 62.79 − 3.22 = 59.57. Since the unit weight of water is 62.4 lb force/ft³,

$$\text{the volume of the water in the container} = \frac{59.57 \text{ lbf}}{62.4 \text{ lbf/ft}^3}$$

$$= 0.955 \text{ ft}^3$$

EXAMPLE 1.3

A container weighs 5 kg when empty, and 87 kg when filled with water. What is the volume of water it can hold?

SOLUTION:

The weight of water in the container = 87 − 5 = 82 kg. By definition 1 kg of water has a volume of one liter, then

$$82 \text{ kg of water} = 82 \text{ liters}$$

Since

$$1000 \text{ liters} = 1 \text{ m}^3$$

Hence

$$82 \text{ liters} = 0.082 \text{ m}^3$$

Density changes due to temperature influence the performance of some artesian wells discharging thermal water. If the hot water flows, its static water level may rise over the wellhead because its lower density causes free flow from the well. If the well is shut off, allowing the water column to cool, the static head may subside below the wellhead. This will mean that pumping will be initially needed until the water column warms up again.

The variation of density according to pressure is assumed to be zero for almost all hydraulic calculations. In other words, water is generally assumed to be incompressible, even though it is about 100 times more compressible than steel. However, in computations relating to shock waves in water, the water hammer, knowledge of the elastic properties of water is essential.

The ratio of the change of pressure to

Table 1-3 Elasticity of Water

Pressure in thousand pounds per sq. in.	Temperature in °F				
	32	68	120	200	300
	E, Elasticity in Thousand psi				
0.015	292	320	332	308	—
1.5	300	330	342	319	248
4.5	317	348	362	338	271
15.0	380	410	426	405	350

$$\Delta p = -K\frac{\Delta V}{V_0}$$

where Δp = pressure change
ΔV = volume change
V_0 = initial volume
K = modulus of compressibility = $\frac{1}{E}$

the corresponding change of volume is called the *bulk modulus of elasticity*. This modulus of elasticity depends on the temperature and pressure. Table 1-3 shows the relationship between elasticity of water, and the ambient temperature and pressure. Note that regardless of the pressure, water reaches its maximum elasticity at 120°F.

SURFACE TENSION, ADHESION, AND CAPILLARITY

It is known that water rises in clays and fine silts to considerable heights. Clay layers become saturated with water even though the free water table exposed to the atmosphere is dozens of feet below. The cause of this *capillary rise* is complex. One of the reasons for capillary rise is the surface tension of the water. *Surface tension* is caused by the bonding of the molecules of the water in three dimensions under normal conditions. On the surface this condition is not satisfied. Hence the surface molecules increase their bonds along the surface, creating a layer of increased molecular attraction which, although only of the magnitude of

one-millionth of a millimeter, has a significant influence on the physical behavior of water in a porous medium. Surface tension is therefore the added cohesion of water molecules on the surface. Its value depends on the temperature and electrolytic content of the water. Table 1-4 shows the relationship between surface tension and temperature in water.

Table 1-4 Surface Tension of Water

Temperature in °C	Surface Tension	
	g/cm	lb/ft
0	0.0756	0.00518
10	0.0742	0.00508
20	0.0727	0.00497
30	0.0711	0.00486
40	0.0695	0.00475

Small amounts of electrolytes added to the water increase its surface tension. Salts dissolved from adjacent soil particles tend to increase the electrolytic content and, hence, the surface tension in groundwater. On the other hand, organic substances like soaps, alcohol, or acids decrease the surface tension. The reducing effect of soap on surface tension make it possible to stretch water film while blowing soap bubbles.

Another important contributor to the physical effect of capillary rise is the *adhesion* of water to most solid materials. Solids that have positive adhesion to water are called hydrophile (water liking), and those that repel water are hydrophobe. The latter have negative adhesion to water. Adhesion between fluids and solids is expressed by the contact angle at the edge of the contacting surfaces, as shown in Figure 1.2a. Hydrophobe materials have a contact angle with water that is larger than 90 degrees. For example, the contact angle between water and paraffin is 107 degrees; hence, paraffin is a good waterproofing agent. Silver, on the other hand, is neutral to pure water; its contact angle is nearly 90 degrees. Quartz and other materials found in porous soils have a contact angle with water that is less than 90 degrees; this means that they wet well by water. The contact angle between ordinary glass and water containing impurities, for example, is about 25 degrees. In fact, the adhesive forces between water and soil particles are so large that they can be separated only by evaporating the water.

The capillary action, the rise of water in the small pores of soils and in thin glass tubes, is caused by the combined action of surface tension and adhesion. Figure 1.2b

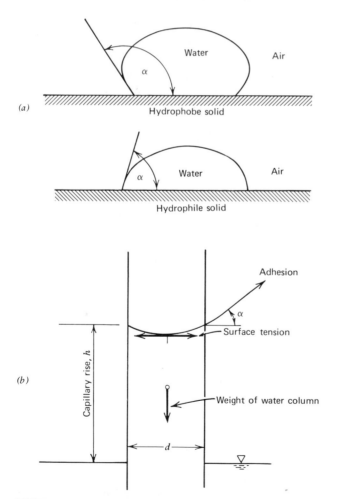

FIGURE 1.2 Causes of capillary action.

depicts the conditions present in a small diameter glass tube in which capillary rise of water takes place. By its adhesion to the solid wall water wants to cover as much solid surface as possible. However, by the effect of the surface tension the water molecules adhering to the solid surface are connected with a surface film in which the stresses cannot exceed the maximum possible surface tension of the water. The molecules in this surface film are joined to molecules below it by cohesive forces. As the adhesion drags the surface film upward, the film then raises a column of water filling the tube upward, against the force of gravity. The outcome of these factors is that the water in the small capillary tube, or in the small pores of soils, will rise upward against the force of gravity to a height at which the ultimate supporting capacity of the surface film is reached. Of course the column of water below the surface film is under tension, which means that the water pressure in a capillary tube is below the atmospheric pressure. The capillary rise is inversely proportional to the diameter of the tube or to the pore size in soils. Hence the finer are the soil grains, the thicker will be the capillary layer in the soil mass. Table 6-2 shows the average heights to which capillary water rises in various types of soils. In the idealized case of a small diameter tube the height of the capillary rise, h, is

$$h = \frac{\pi \sigma \cos \alpha}{d \gamma}$$

in which d represents the diameter of the tube, γ is the unit weight of the water, σ is its surface tension, and α is the contact angle representing the adhesion between the water and the tube.

VISCOSITY

Perhaps the most important physical property of water is its resistance to shear or angular deformation. The measure of resistance of a fluid to such relative motion is called viscosity. We define viscosity as a capacity of a fluid to convert kinetic energy, energy of motion, into heat energy.

The energy converted into heat is considered lost because it can no longer contribute to further motion. It either results in warming up the fluid or is lost into the atmosphere by dissipation. The energy required to move a certain amount of water through a pipe, open channel, or hydraulic structure is determined by the amount of *viscous shear losses* to be encountered along the way. Therefore viscosity of the fluid inherently controls its movement. Viscosity is due to the cohesion between fluid particles and also to the interchange of molecules between the layers of different velocities. Mathematically the relationship between viscous shear stress and viscosity is expressed by Newton's law of viscosity. This is written as

$$\tau = \mu \frac{\Delta v}{\Delta y} \qquad (1.3)$$

which is nothing else but an expression of proportionality between viscous shear resistance and the rate of change of velocity in the direction perpendicular to the shear stress, as shown in Figure 1.3. The mechanism illustrated is in some ways similar to the case of a deck of cards dragged along on a table. The relative velocity between adjacent cards is Δv, the thickness of a card is Δy. The factor of proportionality is called *absolute* (or

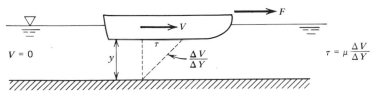

FIGURE 1.3 Interpretation of Newton's law of viscosity.

Table 1-5   Viscosity of Water

Temperature in °C	Absolute Viscosity	
	Centipoise	lb sec/ft^2
0	1.792	0.374×10^{-4}
4	1.567	0.327×10^{-4}
10	1.308	0.272×10^{-4}
20	1.005	0.209×10^{-4}
20.2	1.000	0.208×10^{-4}
30	0.8807	0.183×10^{-4}
50	0.5494	0.114×10^{-4}
70	0.4061	0.084×10^{-4}
100	0.2838	0.059×10^{-4}
150	0.184	0.038×10^{-4}

dynamic) *viscosity*, μ, and has the dimension of force per area (stress) times time interval considered. In science viscosity is usually measured in centipoises. Conveniently the absolute viscosity of water equals one centipoise at 20.2°C, which is about room temperature. This fact allows us to use the absolute viscosity of water as a relative standard for viscosities of other fluids. In comparison the absolute viscosity of air is about 0.17 centipoise and that of mercury about 1.7 centipoises. The viscosity of water depends on its temperature (Table 1-5).

One hundred centipoises equal a poise, which is equivalent to one gram per centimeter second. In terms of force it equals 0.1 (newton · second)/meter^2. In the English system of units absolute viscosity is expressed in pounds seconds/ft^2. The conversion is done by dividing centipoises by 478. Table 1-5 also shows absolute viscosities of water in English units.

EXAMPLE 1.4

A raft 3×6 m in size is dragged at a velocity of 1 m/s in a shallow channel 0.1 m deep measured between the raft and the channel bottom. Compute the necessary dragging force, assuming that Equation 1.3 is valid and that the temperature of water is 20°C.

SOLUTION:

From Equation 1.3

$$\tau = \mu \frac{\Delta v}{\Delta y}$$

Assuming that the velocity changes linearly from zero at the channel to 1 m/s at the raft and from Table 1-5, μ at 20°C = 1.005 centipoises, 1 poise = 100 centipoises; therefore 1.005 centipoises = 0.01005 poise. But

$$1 \text{ poise} = \frac{0.1 \text{ N} \cdot \text{s}}{\text{m}^2}$$

Hence

$$\mu = 1.005 \times 10^{-3} \frac{\text{N} \cdot \text{s}}{\text{m}^2}$$

and

$$\tau = 1.005 \times 10^{-3} \frac{\text{N} \cdot \text{s}}{\text{m}^2} \frac{10 \text{ m/s}}{0.1 \text{ m}}$$

$$= 10.05 \times 10^{-3} \frac{\text{N}}{\text{m}^2}$$

The dragging force of the $3 \times 6\,m^2$ raft is then

$$F = \tau \cdot A = (10.05 \times 10^{-3})18 = 0.1809\ N$$

In English units $1\ lb = 4.448\ N$; hence the dragging force is

$$F = 0.04067\ lb\ of\ force$$

Dividing the absolute viscosity by the density of the fluid at identical temperature results in the *kinematic viscosity, v*. The concept of kinematic viscosity is often used by scientists and engineers. The unit of kinematic viscosity in the S.I. system is cm²/s, or the Stoke; in the English system of units its equivalent is in ft²/sec. The conversion is performed on the basis that 929 Stokes equal 1 ft²/sec. Table 1-6 lists conversion factors for kinematic viscosity.

Viscosity of water depends little on the pressure, but it varies considerably with the temperature. This phenomenon is often neglected, although in some cases the practical influence is most noticeable. During winter the yield of wells feeding from nearby rivers may decrease by as much as 30%, a significant effect caused by this factor.

Table 1-6 Conversion Factors for Kinematic Viscosity

ν	m²/s	m²/hr	cm²/s	ft²/sec	ft²/hr
m²/s	1	3600	1×10^4	10.7639	3.875×10^4
m²/hr	277.8×10^{-6}	1	2.778	299.9×10^{-4}	10.7639
cm²/s (Stoke)	1×10^4	0.36	1	10.7639×10^{-4}	3.875
ft²/sec	0.092903	334.45	929.03	1	360
ft²/hr	25.806×10^{-6}	0.092903	0.25806	277.8×10^{-6}	1

EXAMPLE 1.5

In pipe flow problems the viscosity of the water is taken into account in a dimensionless term called the Reynolds number, expressed by Equation 4.4 as $\mathbf{R} = VD\rho/\mu$. Compute the Reynolds number for a flow of water at 70°F, in a pipe of 40 in. diameter, and at a velocity of 3.3 fps.

SOLUTION:

The density of water at $70°F = 1.936\ slug/ft^3$.

$$\mu = 2.11 \times 10^{-5}\ lbf\ sec/ft^2$$

Then from

$$\mathbf{R} = \frac{VD\rho}{\mu}$$

$$= \frac{3.3 \times 40/12 \times 1.936}{2.11 \times 10^{-5}} = 10^6$$

EXAMPLE 1.6

The seepage loss from a reservoir of 10,000 ft² surface area and 3 ft mean depth was 3 in. per month during the summer when the average water temperature was 25°C. If we assume that the ground water level was well below the reservoir bottom, such that it does not influence the seepage losses, estimate the seepage losses during the winter months when the average water temperature is 4°C.

SOLUTION:

The seepage out of a reservoir depends on the permeability of the soil on the bottom of the reservoir. Equation 6.3 shows that the permeability is proportional to the unit weight of the water and inversely proportional to the viscosity of the water. From Table 1-2 the increase of density of water from 25°C to 4°C is only 3% (from 0.9971 to 1.00 gram/cm³), a negligible quantity. From Table 1-5 the increase of viscosity from 25°C to 4°C is

$$\frac{(1.567 - 0.940) \times 100}{0.94} = 66.67\%$$

Accordingly the permeability and therefore the seepage losses will decrease during the winter by the factor of 0.67. Hence the monthly seepage loss in the winter will be

$$(1 - 0.67) \times 3 \text{ in.} = 1 \text{ in.}$$

PROBLEMS

1.1   How much heat must be removed from 15 m³ of water in order to freeze it, if the initial temperature of the water is 12°C?

1.2   Determine the minimum allowable pressure at the suction side of a pump moving water at 40°F, if the pump is located at a site where the atmospheric pressure is only 90% of the standard atmospheric pressure? (Standard atmospheric pressure is defined as 2116 lb/ft², or 34 ft of water, or 14.7 psi, or 10 m of water.)

1.3   Express the difference in weight between 10 gallons of water at 35 and 85°F.

1.4   Determine the reduction of capillary rise if the contact angle representing the adhesion between water and soil material is increased from 25 to 60 degrees.

1.5   Prove that the unit of absolute viscosity expressed in English units (lb sec/ft²) equals 478 poises.

1.6   Calculate the kinematic viscosity of water at 80°C, expressing it in both English and metric units.

1.7   By using Equation 6.3 showing Darcy's permeability coefficient, determine the change of permeability if the temperature of the water rises from 20 to 80°C.

1.8   Determine the error in the discharge measurement by an elbow meter that was calibrated at 30°C, if the water flowing in the elbow is at 4°C. (See Equation 10.13.)

1.9   What will be the change in a Reynolds number (Equation 4.4) if the temperature of the fluid changes from 40 to 110°F?

2

LAWS OF FLUID MECHANICS

Knowledge of the basic physical laws controlling the motion of all fluids is essential for the understanding of the behavior of water. These fundamental principles will be reviewed in this chapter. In all later chapters we will frequently refer to them.

FLOW RATE

In the science of fluid mechanics we distinguish between compressible and incompressible fluids. Laws relating to the behavior of compressible fluids like air, steam, or gases fall within the domain of aerodynamics, or gas dynamics. Because for all pressures encountered in hydraulic applications, water in its liquid form retains a constant volume for a given amount of mass, it is considered to be incompressible. The assumption of incompressibility results in a considerable simplification of the fundamental fluid mechanics principles. It allows us to measure amounts of water in volumetric terms instead of in mass.

Discharge, meaning volume of water flowing through a certain cross-sectional area during a specified time period is usually measured in cubic feet per second,

or cubic meters per second. In some specific applications other discharge units may be found. For example, discharge of pumps is quoted in gallons per minute, output of water treatment plants may be given in million gallons per day, and so forth. A table of conversion factors including the most frequently used discharge units is included in Appendix II.

By *velocity* we mean the change of position of a water particle in the moving fluid during a certain time interval. In Figure 2.1, for example, a particle may be found at a position x_1 at time t_1. If the same particle at a later time t_2 is found at position x_2, then the velocity of the particle is $(x_2 - x_1)/(t_2 - t_1)$. As we take shorter and shorter time intervals $(t_2 - t_1)$, the path of the particle between the end points x_1 and x_2 will be less and less distinguishable from the straight line connecting the end points. For an extremely small interval (Δt) the

13

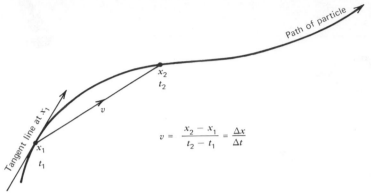

FIGURE 2.1   The concept of velocity vector.

line becomes tangent to the path at point x_1. In this case we speak of a velocity vector at point x_1. The direction of a velocity vector shows which way the particle is moving at this point. The magnitude of the velocity vector signifies how fast the particle is moving. Now, if we are looking at the point x_1 as a location fixed in space through which a stream of fluid particles pass, we can define another important principle: If the succession of water particles passing through x_1 have identical velocity vectors, we speak of *steady flow*. Conversely, if the velocity vector changes for successive fluid particles, we have *unsteady flow*.

In practice we often find that as a discharge Q flows through a cross-sectional area A of a flow channel the velocity of the fluid particles flowing through each point of the area is different.

Usually in the central portion of the cross section we find the highest velocities, while along the edges the velocity may be almost zero. It is convenient in such cases to define the *average velocity* as

$$v_{\text{average}} = \frac{Q}{A} \qquad (2.1)$$

where A is considered perpendicular to the direction of the flow, (Figure 2.2). On the left side of Figure 2.2 the *velocity distribution graph* is depicted. In some topics, such as the measurement of flow by velocity meters, when we speak of velocity we refer to a particular velocity vector in the flow field. In seepage computations, discussed in Chapter 6, we must distinguish between *velocity of seepage*, which refers to the rate of flow of water within the open pores of the soil, and *discharge velocity*, which refers to a particular sur-

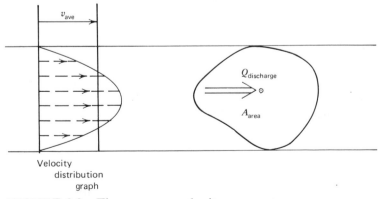

FIGURE 2.2   The average velocity concept.

face area of the porous medium, including its solid portions, as A in Equation 2.1. The two terms are related to each other by the porosity of the medium, which is the ratio of the volume of the voids within the soil particles to the total volume. In our work we will speak of discharge velocity rather than seepage velocity.

Acceleration of a fluid means a change in velocity. There are two kinds of accelerations since the velocity can change in place as well as in time. In a pipe of enlarging diameter, as shown in Figure 2.3,

Internal accelerations within the fluid result from the relative movement of the water because of external forces. By relative movement we mean the motion of water with respect to its surroundings. The external forces may be created by the gravitational acceleration, g, which acts everywhere on Earth. The driving force, in this case, comes about by virtue of Newton's law of motion, defined by Equation 1.1. If the only acceleration acting on the fluid particles is the one due to gravity and the water is contained, we have no cause

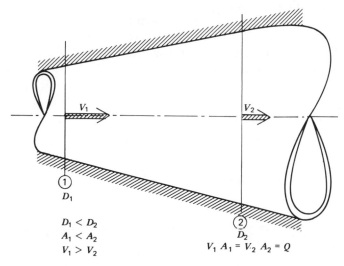

FIGURE 2.3 Continuity relationships without storage.

the velocity of the flow decreases as it passes from a section of small diameter to a section of large diameter. This form of deceleration, or negative acceleration, is called *spatial acceleration*, since the change of velocity occurs in space. If the discharge flowing through a certain cross-sectional area varies with time, we speak of *temporal acceleration*. For almost all topics considered in this book we will assume that no temporal accelerations exist. This means that the flow is steady.

for relative motion between the fluid particles. As long as the water is contained, it will be at rest. Forces caused by water at rest are in the domain of *hydrostatics*, discussed in Chapter 3. Forces caused within the moving mass of fluid, on the other hand, are within the domain of *hydrodynamics*. Hydrodynamics is the study of the behavior of water in motion. Its basic physical laws are the laws of conservation and the laws of energy transfer.

EXAMPLE 2.1

Compute the average velocity of water in a pipe of 12 in. diameter when the discharge of the pipe is 20 gpm.

SOLUTION:

From our table in Appendix II, 1 gpm = 2.227×10^{-3} cubic feet per second (cfs); hence, 20 gpm = 0.0445 cfs. The cross-sectional area of a 12-in., or 1-ft diameter pipe is

$$A = \frac{\pi d^2}{4} = \frac{\pi}{4} 1^2 = 0.785 \text{ ft}^2$$

By Equation 2.1

$$v_{\text{ave}} = \frac{Q}{A} = \frac{0.0445}{0.785} = 0.0567 \text{ fps}$$

EXAMPLE 2.2

A 20 by 20 ft excavation is surrounded by sheetpiles. In order to remove the water entering from the soil on the bottom, a pump is installed that delivers 10 gallons per minute (gpm). The soil is loose sand with a porosity of 42%. Compute the discharge velocity and the seepage velocity of the entering groundwater.

SOLUTION:

The discharge, 10 gpm equals 0.02227 ft³/sec. The superficial area of the flow is $20 \times 20 = 400$ ft². Hence the average discharge velocity is

$$v = \frac{Q}{A} = \frac{0.02227}{400} = 5.57 \times 10^{-5} \text{ fps}$$

which, in metric units, is

$$v = 5.57 \times 10^{-5} \frac{\text{ft}}{\text{sec}} 30.48 \frac{\text{cm}}{\text{ft}}$$

$$= 1.69 \times 10^{-3} \text{ cm/s}$$

The seepage velocity, that is, the average velocity of the groundwater within the pores of the sand is

$$v_{\text{seepage}} = \frac{v_{\text{discharge}}}{\text{porosity \%}} \times 100$$

$$= \frac{1.69 \times 10^{-3} \text{ cm/s}}{42\%} 100$$

$$= 4.24 \times 10^{-3} \text{ cm/s}$$

or

$$= 4.24 \times 10^{-5} \text{ m/s}$$

EXAMPLE 2.3

The discharge of a creek was measured to be 0.5 m³/s at 9 a.m. and 0.6 m³/s at 4 p.m. on the same day. The area of the flow was 1.445 m² during the first measurement and 1.57 during the second measurement. Determine the average velocities of the flow and the average temporal acceleration during this time period.

SOLUTION:

At 9 a.m. the velocity was

$$v_1 = \frac{Q_1}{A_1} = \frac{0.5}{1.445} = 0.346 \text{ m/s}$$

At 4 p.m.

$$v_2 = \frac{Q_2}{A_2} = \frac{0.6}{1.57} = 0.382 \text{ m/s}$$

The time interval between 9 a.m. and 4 p.m. is 7 hours, or

$$T = 7 \times 3600 = 25{,}200 \text{ seconds}$$

The average temporal acceleration

$$a = \frac{v_2 - v_1}{T} = \frac{0.382 - 0.346}{25{,}200} = \frac{0.036}{25{,}200}$$
$$= 1.43 \times 10^{-6} \text{ m/s}^2$$

CONSERVATION OF MASS

Of the three conservation laws of physics one is the conservation of mass. It states that mass cannot be created or destroyed. This concept gives rise to the *equation of continuity*, which states that within any hydraulic system the discharge flowing in, the stored volume within, and the discharge flowing out must be balanced; in other words, all volumetric quantities must be accounted for. Since we consider water to be incompressible, mass or volume may be used interchangeably in this respect. To put the concept into mathematical form we may write our continuity equation as

$$Q_{in} - Q_{out} = \text{change in storage} \quad (2.2a)$$

The equation above is often used in the analysis of reservoirs and in routing of floods in rivers. It is important that the time scale be the same on both sides of the equation. For example, if the discharges are available in the units of "cfs" and the change in storage is desired for periods of, say, six hours an appropriate conversion factor must be introduced.

In the case when no change in storage is possible, as in a pipe flowing full, the right side of Equation 2.2a will reduce to zero, that is,

$$Q_{in} - Q_{out} = 0 \quad (2.2b)$$

which means that what goes in, must come out. Figure 2.4 illustrates these concepts. In certain applications, as in rivers or complex pipelines, the problem is broken into smaller components joined together at certain points. In this case, care should be exercised in selecting the proper sign for each discharge components. A common sign convention is to consider discharges flowing into the hydraulic component considered as positive and outflowing discharges as negative. At points where two hydraulic components are joined the sign of the discharge changes.

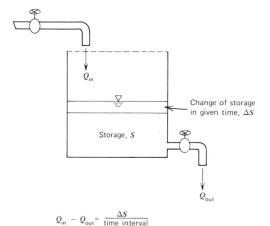

$$Q_{in} - Q_{out} = \frac{\Delta S}{\text{time interval}}$$

FIGURE 2.4 Continuity relationships with storage.

EXAMPLE 2.4

In a farm's water supply system a 10,000 gallon underground storage tank is continually replenished from a well by a submersible pump delivering 2 gpm. Assuming that the reservoir is initially full, how much time will it take to empty out the reservoir by using a 30 gpm pump?

SOLUTION:

By Equation 2.2a

$$30\text{ gpm} - 2\text{ gpm} = 28\text{ gpm} = \text{change in storage}$$

Hence,

$$\frac{10,000\text{ gallons}}{28\text{ gallons/minute}} = 357\text{ minutes} = 5.95 \sim 6\text{ hours}$$

EXAMPLE 2.5

A 12-in. diameter pipe reduces to a diameter of 6 in., then expands to 8 in. If the average velocity in the 6 in. pipe is 16 ft per second (fps), what is the discharge in the pipeline and what are the average velocities in the other two pipes?

SOLUTION:

The cross-sectional area of pipe is

$$A = \frac{\pi}{4} D^2$$

For a 6 in. pipe

$$A_6 = \frac{\pi}{4} \left(\frac{6}{12}\right)^2 = 0.196\text{ ft}^2$$

For an 8 in. pipe

$$A_8 = \frac{\pi}{4} \left(\frac{8}{12}\right)^2 = 0.349\text{ ft}^2$$

And for a 12 in. pipe

$$A_{12} = \frac{\pi}{4} \left(\frac{12}{12}\right)^2 = 0.786\text{ ft}^2$$

From the Equation 2.1

$$Q = v \cdot A$$

Since the discharge in the pipeline is constant,

$$Q = v_6 A_6 = 16\frac{\text{ft}}{\text{sec}} (0.196\text{ ft}^2) = 3.14\text{ cfs}$$

Then the velocities in the other two pipes are

$$v_8 = \frac{Q}{A_8} = \frac{3.14}{0.349} = 9.0\text{ fps}$$

$$v_{12} = \frac{Q}{A_{12}} = \frac{3.14}{0.786} = 4.0\text{ fps}$$

CONSERVATION OF MOMENTUM

Another conservation law is the conservation of momentum. *Momentum* of a flowing fluid is the product of its mass flow rate (ρQ) and its velocity. Momentum may be considered as a property of the flowing fluid:

$$M = \rho Q v = \rho A v^2 \qquad (2.3a)$$

where A is the cross-sectional area of the flow channel and v is the average velocity of the flow. Flow of water in a pipe of D diameter, for example, has a momentum of

$$M = 4\rho \frac{Q^2}{\pi D^2} \qquad (2.3b)$$

since the average velocity in a pipe is

$$v = \frac{4Q}{\pi D^2} \qquad (2.4)$$

and ρ is the density of the fluid.

Change of momentum along the path of a flowing fluid results in a force called *impulse force, F.* For steady flows the discharge Q is constant; hence, the time rate of change of the velocity v is zero. Therefore, for steady flows, changes in momentum may be caused only by spatial changes in velocity. Between two points (points 1 and 2 shown in Figure 2.5) along the path of the flow the change of momentum can be written as

$$\begin{aligned} M &= M_2 - M_1 \\ &= \rho Q v_2 - \rho Q v_1 \\ &= \rho Q (v_2 - v_1) \\ &= F \end{aligned} \qquad (2.5)$$

One must keep in mind that the velocities, as well as the impulse force, are vector quantities. Therefore these terms may change either in magnitude, in direction, or in both. Equation 2.5 embodies the *law of momentum conservation*, which states that momentum may not be lost in a hydraulic system, although some of it may be converted into impulse forces. Hence, if the mass flow rate and the physical configuration of the flow channel causing a change in flow direction are known, the resulting impulse forces acting on the hydraulic component or structure may be computed. The same concept may be used to measure the discharge, as done for the measuring elbow (Equation 10.13) in which the discharge, and consequently the velocity, is unknown and determined by measuring pressures generated by the impulse force acting on the pipe elbow.

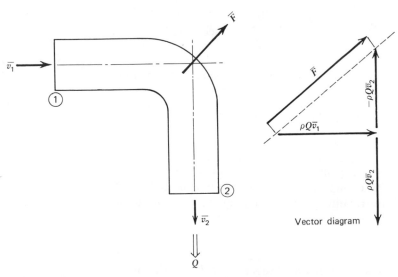

FIGURE 2.5   Impulse force on a pipe elbow.

EXAMPLE 2.6

The rate of discharge from a fire hose nozzle is said to be expressible by the formula

$$Q = 27d^2p^{0.5}$$

in which Q is in gpm, d is the internal diameter of the nozzle in inches, and p is the pressure in the fire hydrant in psi. Determine the impulse force on a perpendicular plane caused by a jet of water through a 2-in. nozzle if the pressure in the hydrant is 100 psi.

SOLUTION:

Using the equation above we first determine the discharge:

$$Q = 27 \times 4 \times 10 = 1080 \text{ gpm}$$
$$= 2.227 \times 10^{-3} = 1080 = 2.4 \text{ cfs}$$

The velocity in the 2-in. nozzle is

$$v = \frac{Q}{A} = \frac{2.4}{\pi \left(\frac{2}{12}\right)^2 \Big/ 4} = 110 \text{ fps}$$

The density of water may be computed from Equation 1.2 or may be taken from Table 1-2:

$$\rho = \frac{\gamma}{g} = \frac{62.4}{32.2} = 1.937 \frac{\text{lbf sec}^2}{\text{ft}^4}$$

The mass flow rate in this case is then

$$\rho Q = 1.937 \times 2.4 = 4.65 \frac{\text{lbf sec}}{\text{ft}}$$

The jet of water contains a momentum that may be computed from Equation 2.3a as the product of the mass flow rate and the velocity,

$$M = (\rho Q)v = 4.65 \times 110 = 511.37 \left[\frac{\text{lb sec}^2}{\text{ft}^4} \cdot \frac{\text{ft}^3}{\text{sec}} \cdot \frac{\text{ft}}{\text{sec}}\right]$$
$$= 511 \text{ lb}$$

When the jet hits a perpendicular immobile plane, all of this momentum converts into an impulse force; hence,

$$F = M = 511 \text{ lb of force}$$

EXAMPLE 2.7

A 90° elbow in a 6-in. pipeline carries 3 cfs of water at room temperature (Fig. E2.7). Compute the magnitude and direction of force acting on the elbow resulting from the change of momentum of the flow.

SOLUTION:

The momentum of flow entering the elbow is ρQv. From Equation 2.4

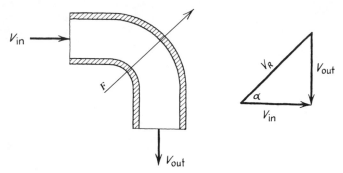

FIGURE E2.7

$$v = \frac{Q}{A} = \frac{4Q}{\pi D^2} = \frac{4 \times 3}{\pi \left(\frac{6}{12}\right)^2} = 15.28 \text{ fps}$$

The direction of the resultant force vector is α,

$$\text{where } \alpha = \tan^{-1}\left(-\frac{v_{out}}{v_{in}}\right) = \tan^{-1}(-1) = 45°$$

The magnitude of this velocity change

$$= \mathbf{V}_R = \mathbf{V}_{in} - \mathbf{V}_{out}$$

and

$$\cos \alpha = \frac{v_{in}}{v_R}$$

$$v_R = \frac{v_{in}}{\cos(45°)}$$

$$= \frac{15.28}{0.707} = 21.61 \text{ fps}$$

From Equation 2.5

$$F = \rho Q(v_2 - v_1)$$
$$= \rho Q v_R$$
$$= \left(\frac{62.4}{32.2}\right)(3)(21.61)\left[\frac{\text{lbf sec}^2}{\text{ft}^4} \cdot \frac{\text{ft}^3}{\text{sec}} \cdot \frac{\text{ft}}{\text{sec}}\right]$$
$$= 125.63 \text{ lbf}$$

is the magnitude of the force acting on the elbow due to the flow of water.

The force due to momentum changes on the blades of a turbine may cause the turbine to rotate. The product of an impulse force and the radius of rotation results in the *moment of momentum* or, as it is also called, the *torque*. The concept of torque is applied in Equation 5.3 to deter- mine the power requirements of pumps. Figure 2.6 helps to explain the concept of torque. The impulse force transferred to the water by the ship's propeller causes the ship to move. This force must be sufficient to overcome the various losses due to this relative motion. One of the

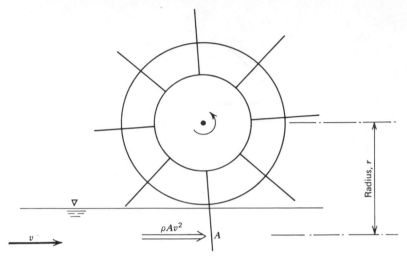

FIGURE 2.6   Torque of water on a water wheel.

losses among others is the momentum needed to be given to the water first to move out of the way of the ship then to move back behind it. This process causes relative movements between the water particles around the ship that give rise to additional losses due to the resistance by the viscosity of the water.

EXAMPLE 2.8

From a 12 in. diameter garden sprinkler two horizontal jets of water are issued in a tangential direction. The velocity of the jets is 30 fps, and the discharge of the sprinkler is 0.5 cfs. Compute the torque generated by the sprinkler that causes its rotation.

SOLUTION:

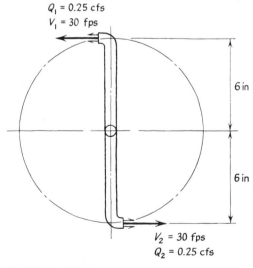

FIGURE E2.8

Since the total discharge = 0.5 cfs, the discharge of each jet = 0.5/2 = 0.25 cfs. Then the tangential impulse force from one arm is the product of the discharge, density, and the velocity of the jet as given in Equation 2.3:

$$F = \rho Q v = \frac{62.4}{32.2}(0.25)(30) = 14.53 \text{ lbf}$$

This force is tangent to the radius of motion, which is 0.5 ft. Therefore the torque from each arm

$$= 14.53 \times 0.5 = 7.265 \text{ ft-lbf}$$

Hence the total torque

$$= 7.265 \times 2 = 14.53 \text{ ft-lbf}$$

WORK, ENERGY, AND POWER

Force exerted over a distance represents *work* done. For example, to push a certain amount of water through a pipeline requires that work be done. Work, therefore, is defined as

$$W = F \cdot l \qquad (2.6)$$

in which F is the force acting in the direction of l distance.

Power is a term closely related to work. It refers to work done per unit of time. To express power mathematically, we write

$$P = \frac{W}{t} \qquad (2.7a)$$

where t is the time required to do the work. To perform a certain amount of work in a shorter period of time requires more power than to perform the same amount of work over a longer period of time. We use the concept of power in conjunction with pumps, as in Equation 5.2. Power required of electric motors driving pumps is most often expressed in *horsepowers* (*HP*). One horsepower equals 550 foot-pounds per second in the English system of measures. In the metric system the horsepower is only slightly different. One English horsepower equals 76.05 meter-kilograms per second. In terms of electric power we speak of kilowatts. One kilowatt is 1.34 horsepower. To move a discharge Q against a pressure of (γH), the horsepower requirement is

$$HP = \frac{\gamma Q H}{550} \qquad (2.7b)$$

assuming 100% efficiency in the operation.

Equation 2.7b requires that we substitute our variables in terms of pounds, feet, and seconds. In the metric system we use the relationship on the basis that one horsepower equals 75 meter-kilograms per second. Hence, in metric units

$$HP_{\text{metric}} = \frac{\gamma Q H}{75} \qquad (2.7c)$$

in which, of course, all units are in kilograms, meters, and seconds.

For completeness, we note that one horsepower equals 2545 BTU per hour. Hence a pump using one horsepower for a period of one hour overcomes a heat loss of 2545 BTU, neglecting other effects.

EXAMPLE 2.9

A mountain creek discharges 100 cfs of water along a rocky channel without apparent freezing in subzero weather. The flow is uniform in cross section, the average velocity is constant and equals 10 fps. The channel elevation above mean sea level is reduced by 250 ft over a length of 2000 ft. Compute the heat energy generated by the flow that maintains an above-freezing temperature in the water.

SOLUTION:

As the velocity is constant along the channel, then the kinetic energy is also constant. The flow is exposed to the atmosphere. Hence, no pressure energy is present so that the energy gained over the 2000 ft length is due to gravity only.

From Equation 2.9

$$h_L = \left(\frac{v_1^2}{2g} - \frac{v_2^2}{2g}\right) + \left(\frac{p_1}{\gamma} - \frac{p_2}{\gamma}\right) + (z_1 - z_2)$$

$$= 250 \left(\frac{\text{ft-lbf}}{\text{lbm}}\right) = \text{energy gained from flowing.}$$

The total energy gained $= mh_L \left(\frac{\text{ft-lbf}}{\text{lbm}} \cdot \frac{\text{lbm}}{\text{sec}}\right)$ where $m = \rho Q$

$$= \frac{(100)(62.4)(250)}{32.2} \left(\frac{\text{ft-lbf}}{\text{sec}}\right)$$

$$= 48{,}447 \left(\frac{\text{ft-lbf}}{\text{sec}}\right)$$

As $1 \left(\frac{\text{ft-lbf}}{\text{sec}}\right) = 0.07717 \left(\frac{\text{BTU}}{\text{min}}\right)$

Then the total gained heat energy $= 3738 \dfrac{\text{BTU}}{\text{min}}$

Also 1(ft-lbf/sec) = 1.356×10^{-3} kilowatts.
Hence $(48{,}447)1.356 \times 10^{-3} = 65.7$ kW.
This is equivalent to the heat of a 100 W light bulb for each 3 ft length of the channel.

The capacity to do work is called *energy*. Energy therefore is simply stored work. In fluid mechanics, energy is most often expressed as energy per unit mass of fluid. There are various forms in which energy appears in nature; put another way, one can store work in several ways. Temperature indicates the amount of *heat energy* a substance contains. If we know the specific heat of a substance, we may compute the amount of heat energy that could be removed from it by reducing its temperature. We have learned in Chapter 1 that temperature and pressure represent internal energy in matter. A lump of coal has *chemical energy* that may be transformed into heat energy by the chemical reaction called burning. An object placed higher in a gravitational field has more energy by virtue of its relative elevation than the same object at a lower elevation. This is called *potential energy*. A compressed spring or a stretched strip of rubber has *elastic energy*. The force applied to them gives rise to *pressure energy*. Pressure energy and elastic energy may be illustrated by considering an inflated bicycle tube; the elastic rubber tube is stretched, gaining elastic energy; since the air in the tube is under pressure, it is compressed and has pressure energy with respect to the surrounding air, which itself is under pressure by the atmosphere. Moving objects contain *kinetic energy.* Their velocity may propel them to higher elevation, like water from a fire hose, in which case the kinetic energy is converted into potential energy.

As we have seen above, some energies can be freely converted among each other; others are not so reversible. Elevation or potential energy, pressure energy, and kinetic energy are reversible energies. These are called hydraulic energies. On the other hand, *heat energy* cannot be harnessed by hydraulic means. Once a portion of our hydraulic energy converts into heat through viscous shear action, it is lost to us; therefore, we refer to it as energy loss. Yet the heat energy converted from sunshine causes the water from the oceans to evaporate, keeping water in constant circulation through creeks and rivers.

CONSERVATION OF ENERGY

An important concept of physics is the law of energy conservation. It states that energy cannot be lost, though it may be converted into other forms. In other words, the theorem states that in a hydraulic system the sum of all energies is a constant. Writing this total energy as E in foot-pounds of force per pounds of mass, we can summate the various forms in which energy may appear as

$$\text{kinetic energy} = v^2/2g$$

$$\text{pressure energy} = p/\gamma$$

$$\text{elevation energy} = y$$

$$\text{total hydraulic energy} = E$$

or in a mathematical form

$$E = \frac{v^2}{2g} + \frac{p}{\gamma} + y = \text{constant} \quad (2.8)$$

which is called *Bernoulli's equation* after one of the famous hydraulicians. The

above listing of energies includes only hydraulic energies. It does not include the heat energy term. Heat energy is always present although the science of hydraulics is not concerned directly with it. Only the change in the heat energy appears in the equations. Heat energy changes appear as water flows from one point to another in the form of viscous shear losses, taking away some portion of the total hydraulic energy E. For this reason, E is called the *total available hydraulic energy* at any point considered. As this total available energy decreases along the direction of the flowing water between two points, as shown in Figure 2.7, the law of energy conservation has this form:

$$\frac{v_1^2}{2g} + \frac{p_1}{\gamma} + y_1 = \frac{v_2^2}{2g} + \frac{p_2}{\gamma} + y_2 + h_{\text{Loss}} \quad (2.9)$$

in which subscripts 1 and 2 refer to points shown in the flow field and the term h_{Loss} refers to the energy lost in the form of heat. This lost heat energy is actually still present in the environment in the form of an imperceptible temperature rise that is generally dissipated in the surrounding atmosphere. The h_L term is commonly called head loss because it is the apparent reduction of the height of the total available hydraulic energy E as the water moved from one point to the other.

The terms in Equation 2.8 all have the dimension of elevation, measured above a horizontal reference level, called *datum plane*, which is perpendicular to the direction of the gravitational acceleration. The sum of the last two terms $p/\gamma + y$, is sometimes called piezometric height. This is the height to which water would rise in a pipe with one of its ends inserted into an arbitrary point of the flow field. The line, shown in Figure 2.7, connecting several points of such piezometric measurements along the path of flow, is called *hydraulic grade line*. It is always below the *total energy grade line* by an amount equal to the kinetic energy head, $v^2/2g$, at the point of the piezometric measurement. In open channels, when the water surface is exposed to the atmosphere and hence it is not under pressure, the hydraulic grade line coincides with the water surface.

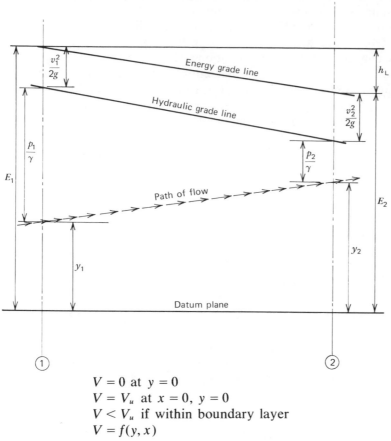

$$V = 0 \text{ at } y = 0$$
$$V = V_u \text{ at } x = 0, \ y = 0$$
$$V < V_u \text{ if within boundary layer}$$
$$V = f(y, x)$$

FIGURE 2.7   Interpretation of the energy equation between two points.

EXAMPLE 2.10

The velocity in a pipe of 4 in. diameter carries 0.025 m³/sec. The pipe is 12 m above the reference level. The pressure in the pipe is measured to be 750 kg/m². Calculate the total hydraulic energy in the pipe at the point in question.

SOLUTION:

The pipe 4 in. diameter = $4 \times 2.54 = 10.16$ cm

The cross-sectional area = $\dfrac{\pi D^2}{4} = \dfrac{\pi}{4}(10.16)^2 = 81.0$ cm²

$$= 81.0 \times 10^{-4} \, \text{m}^2$$

From Equation 2.1

$$v = \frac{Q}{A}$$

$$Q = 0.025 \, \text{m}^3/\text{s} \qquad A = 81.0 \times 10^{-4} \, \text{m}^2$$

Then the velocity $V = \dfrac{0.025}{81.0 \times 10^{-4}} = 3.083$ m/s

the kinetic energy $= \dfrac{v^2}{2g} = \dfrac{(3.083)^2}{2(9.81)} = 0.4844$ m

the pressure energy $= \dfrac{P}{\gamma} = \dfrac{750}{1000} = 0.75$ m

From Equation 2.8 the hydraulic energy

$$E = \frac{v^2}{2g} + \frac{P}{\gamma} + z$$

$$= 0.4844 + 0.75 + 12 = 13.23 \text{ m}$$

This energy is measured at the point 12 m above the reference level.

EXAMPLE 2.11

A 1000 m long pipe is 30 m higher at the entrance point and 10 m higher at the exit point than the reference level (Fig. E2.11). The

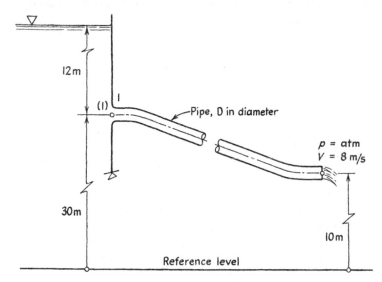

FIGURE E2.11

pipe diameter is constant. The velocity in the pipe is 8 m/s. The water elevation at the entrance is 12 m above the pipe. The pressure at the exit point is the atmospheric pressure. Calculate the energy loss due to the flow in the pipe.

SOLUTION:

The pipe has a constant diameter. The kinetic energy is

$$\frac{v^2}{2g} = \frac{v_1^2}{2g} = \frac{v_2^2}{2g}$$

$$= \frac{(8)^2}{2 \times 9.81} = 3.26 \text{ m}$$

From Equation 2.9, if we take the points at the entrance and the exit point as (1) and (2) we have, therefore,

$$\frac{v_1^2}{2g} + \frac{p_1}{\gamma} + y_1 = \frac{v_2^2}{2g} + \frac{p_2}{\gamma} + y_2 + h_L$$
$$3.26 + 12 + 30 = 3.26 + 0 + 10 + h_L$$

Hence, $h_L = 32$ m = hydraulic energy loss in the pipe due to friction and other factors.

THE CONCEPTS OF ENERGY TRANSFER

The science of hydraulics is primarily concerned with the determination of the magnitude of the energy loss, h_L, under various circumstances. In most practical applications, as in the flow of water in pipes and in open channels, the energy loss is assumed to be proportional to the square of the average velocity of the flow, or in other words, the kinetic energy. Seepage is one exception, where the energy loss is considered to be proportional to the velocity. This concept of proportionality gives rise to coefficients or *factors of proportionality*. This is the reason why hydraulics is a science of coefficients that are determined experimentally. Formulas for pipe flow include a coefficient f, which is called the friction factor; open channel flow formulas include a coefficient n, which is called roughness factor; seepage has its permeability coefficient k; etc. All of these coefficients enable us to determine the rate of conversion of hydraulic energy into heat energy. This is the central question of hydraulics, but also it is the least understood one. Because of the complexity of the physical mechanism involved, we may not necessarily assume that the influence of various parameters on the rate of conversion is in linear relationship with certain major variables of hydraulic systems. Indeed, since in many instances the relationships cannot be represented by a proportionality constant, these factors are themselves dependent on other variables.

This gives credence to the remark made by Theodore von Kármán, who referred to hydraulics as "the science of variable constants." Still, because of the complexity of the physical mechanisms involved in most hydraulic applications, the influence of various parameters on the rate of conversion of hydraulic energy into heat energy is taken into account by empirical coefficients.

Many generations of hydraulicians strived to find better means to compute the rate of this energy transfer. More mathematical and less empirical methods would give better values than the empirical coefficients, which in many instances are only slightly better than educated guesses.

A major scientific breakthrough came early in our century when Ludwig Prandtl developed his *boundary layer theory*. The boundary layer theory allows engineers to write their theoretical formulas for fluid flow in a manner better representing the physical conditions of nature. The essence of the boundary layer concept is that fluids stick to solid boundaries with the consequence that the velocity of the fluids flowing adjacent to a solid boundary gradually increases from zero at the boundary to a value at which the presence of the nearby solid boundary is not felt. In the region where this gradual increase takes place, as depicted in Figure 2.8, the velocity of the fluid overcomes a retardation due to viscous and turbulent shear effects. This zone is called the boundary layer. For aeronautical applications this boundary layer is very thin as compared to the whole mass of air, but in the flow of

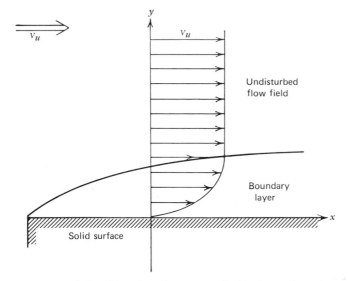

FIGURE 2.8   The development of the boundary layer on a solid surface.

water in pipes and open channels all the flowing fluid is influenced by the pipe or channel surface—that is, the whole flow is dominated by the boundary layer.

There are various formulas derived for generally the simplest cases of physical boundaries on the basis of the classical boundary layer theory. These give a better description for the velocity distribution curve than available previously. Many of these equations, such as the one for pipe flow, Equation 4.7, still contain some empirical coefficients as they were adjusted to better fit the experimental results available. But there are relatively few formulas that were derived for hydraulic applications on the basis of the boundary layer theory. The reason for this is the immense mathematical complexity of the theoretical formulas coupled with the physical complexity of most hydraulic problems.

One of the intrinsic laws of energy transfer is the *law of optimum energy*. In popular terms it states that "the water seeks the path of least resistance." Scientifically this means that of all possible flow distributions the one and only one that may occur in nature is the one requiring the least amount of energy per mass of water, that is, the one delivering the most fluid under the given conditions of energy availability. This law controls flows in pipe networks, open channels, and seepage, to mention only a few of the important situations. Solutions obtained by statistical methods using the so called *probabilistic approach* indicate that the correct answer that satisfies this minimum energy theorem corresponds to the mean value of all possible answers. Most common methods of hydraulic analysis satisfy this basic law intrinsically. Where a minimization theorem is to be specifically stated, the computations are complex beyond the scope of our study.

PROBLEMS

2.1   In a 4-ft diameter pipe the velocity distribution is semicircular with its maximum value being 3 ft/sec at the centerline. The velocity at the pipe wall is zero. Calculate the discharge and the average velocity in the pipe.

2.2   A storage reservoir is continuously filled through a pipe delivering 2 cfs. The reservoir is circular in shape with a diameter of 20 ft. The height of the tank is 12 ft. In 15-minute intervals a pump operates for 5 minutes draining out 300 ft³ of water. Determine the time it takes for the tank to overflow, assuming that it is initially empty.

2.3   The discharge in a 2-ft diameter pipe is 4 cfs of water. Determine the momentum of the flow.

2.4   A 50 liters/s discharge is flowing in a 25 cm diameter, 180 degree pipe elbow. Determine the magnitude and direction of the impulse force acting on the fixture.

2.5   A water wheel delivers a torque of 1250 lb-ft at 30 RPM. The radial distance of the point of impact from the axis of the wheel is 5 ft. If the submerged area of one blade of the wheel is 15 ft², determine the average velocity of the water.

2.6   A pump is delivering 50 cfs of water against a total head of 30 ft. How much is the theoretical power requirement in terms of horsepower and kilowatts?

2.7   The bottom of a rectangular flume is 25 m above reference level. The depth of the water in the flume is 2 m. The average velocity is 0.5 m/s. Compute the total hydraulic energy at the surface of the water and at the bottom of the flume.

2.8   A 1200-ft long pipeline has, at its inlet point, 155 ft of available hydraulic energy with respect to the reference level. The velocity in the pipe is 3 fps; the pressure at its end is 15 psi. The elevation of the end of the pipe is 10 ft above reference level. How much energy was lost by the water while flowing through the pipe?

2.9   How much heat is dissipated per unit length of pipe and per second due to the energy loss in the flow described in Problem 2.8. (Express results in BTUs.)

2.10   Calculate the slope of the total energy grade line and the hydraulic grade line for Problem 2.8. Determine the piezometric height at a point in the pipe 400 ft from its entrance.

2.11   By the aid of a pump a 12-in. diameter pipe delivers 16 cfs of water over a 2000 ft long distance. The entrance of the pipe is at 862 ft, the exit is at 1020 ft above reference level. How much initial energy is required if the energy loss in the pipe is 120 ft and the pressure required at the exit point equals 25 ft of water?

3

WATER PRESSURE

The magnitude and direction of hydrostatic forces acting on all structures and devices constructed to retain or contain water are important elements in the structural design of those structures. In this chapter the determination of these hydrostatic forces is explained for plane and curved boundary surfaces as well as for floating bodies.

HYDROSTATIC PRESSURE AND FORCE

Structures constructed for the purpose of retaining water are subjected to hydrostatic forces. Hydrostatic pressure is the manifestation of pressure energy in water expressed by

$$E_p = \frac{p}{\gamma} + y \qquad (3.1)$$

in which γ is the unit weight of water as shown in Table 1-2. The term E_p is represented by the depth y of the water over the point at which p is sought. The depth y is to be measured always in the direction of the gravitational acceleration.

Since water is a fluid, it cannot resist shear forces as solid materials do. For this reason the pressure p of water at a point in question always acts in all directions.

Therefore, if one wishes to represent the water pressure at a depth y in the form of a vector, the result is a sphere of radius p around the point where the pressure is acting. For the same reason when the pressure is evaluated along the containment boundaries of the water body, the direction of the pressure vector on the boundary must always be perpendicular to the boundary surface. Also, it follows from Equation 3.1 that the total amount of water present has no bearing on the magnitude of the pressure, since it depends only on the depth below the piezometric surface.

If hydrostatic pressure acts on an elementary, small surface area ΔA, an elementary hydrostatic force ΔF develops that may be computed by the formula

$$\Delta F = p \cdot \Delta A = \gamma \cdot y \cdot \Delta A \qquad (3.2)$$

31

Since water cannot support shear forces without motion (see Equation 1.3) the hydrostatic force must be perpendicular to the surface area on which it acts. Hence the orientation of the area ΔA in Equation 3.2 determines the direction of the elemental hydrostatic force ΔF that acts over it.

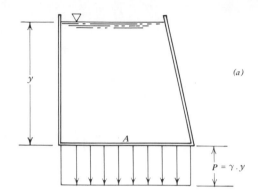

(a)

The determination of the total hydrostatic force F acting on a surface area A is done by summarizing all elementary force components acting on all small ΔA surface components. The resultant hydrostatic force could be considered as acting at a single point of the surface. This point on the surface is called the action point of the total hydrostatic force, F.

In using the English system of measurement the force is expressed as pounds, the unit weight γ being a gravitational term, lb/ft³. In the S.I. system we speak of water density. In this case Equation 3.2 results in kilograms, which is the mass weight. As long as this distinction is understood Equation 3.2 may be used in both systems of units.

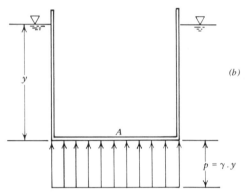

(b)

Depending on the orientation of the surface area, computations of hydrostatic forces are more or less simple. In the case of a horizontal plane, the water pressure is constant throughout, and is given by the product of the unit weight and its depth below the surface. The orientation of the force depends on which side of the surface the water is located. Figure 3.1 shows two such cases. In one case the water is above the surface; hence the force is

FIGURE 3.1 Water pressure on vertical planes.

acting downward. In the other case the water is below the surface; therefore the force acts upward. In both cases the point of action is at the center of the area. Another method giving the same result is to determine the pressure at the centroid of the area A and multiplying this pressure with A.

EXAMPLE 3.1

A 3 ft diameter manhole cover is to be placed on a section of a sewer line in which the water flows occasionally under pressure. Compute the pressure of the water at which the manhole cover is lifted if the weight of the cover is 85 lb.

SOLUTION:

The area of the manhole cover $\quad \dfrac{\pi}{4} d^2 = \dfrac{\pi}{4}(3 \times 12)^2$

$$= 1018 \text{ in.}^2$$

Therefore the weight of cover per unit area $= 85/1018 = 0.0835$ lb/in.^{2}. This means that the manhole cover will be lifted if the water pressure exceeds 0.0835 psi, or 12.024 lb/ft^2, which equals 0.192 ft of water.

VERTICAL AND SLANTED PLANES

If a plane surface area is not horizontal the pressure over it will vary according to the variation of the depth. There are two ways to determine the resultant hydrostatic force. A pressure diagram may be constructed by computing pressures at the deepest and at the shallowest points of the area. As the pressure varies linearly with the depth, these two points may be connected by a straight line giving the pres-

sure distribution (Figure 3.2a). The same result is obtained if the pressure is first separated into its horizontal and vertical components, as shown in Figure 3.2b. The magnitude of the force obtained is a sum of all elemental components of the surface area multiplied by the pressure on these elemental areas. In Figure 3.2 we see that the hydrostatic force will not act at the centroid of area A because the pressure is not evenly distributed over it. The point of action is always below the centroid of the area at a distance e, de-

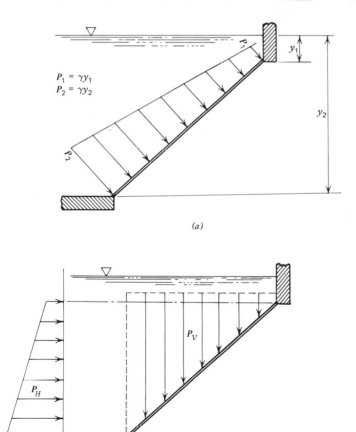

$P_1 = \gamma y_1$
$P_2 = \gamma y_2$

(a)

(b)

FIGURE 3.2 Pressure on slanted planes.

pending on the geometry of the problem. The distance e depends on the shape of the surface in question and its position with respect to the vertical coordinate. Two important concepts of mechanics must be known before the location of the point of action can be determined. These are the first and second moments of areas.

The *first moment of a plane area*, with respect to a coordinate line in the same plane, is computed by multiplying the magnitude of the area A with the perpendicular distance l between its centroid and the coordinate line, as shown in Figure 3.3, that is,

$$M = l \cdot A \qquad (3.3)$$

The *second moment of an area*, also called moment of inertia, is somewhat more complicated. If we are to determine

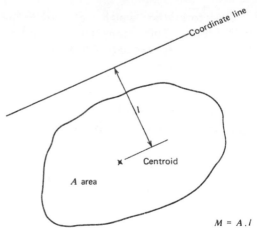

FIGURE 3.3 First moment of an area.

I_0, the second moment of an area with respect to a coordinate line crossing its centroid, we have to sum the products of

Table 3-1 Location of Centroid, Area, and Moment of Inertia of Common Shapes

Rectangle:	$A = b \cdot h$	$I_0 = b \cdot h^3/12$
Triangle:	$A = b \cdot h/2$	$I_0 = b \cdot h^3/36$
Circle:	$A = \pi D^2/4$	$I_0 = \pi R^4/4$
Semicircle:	$A = \pi R^2/2$	$I_0 = 0.110R^4$
Ellipse:	$A = \pi b \cdot h/4$	$I_0 = \pi b \cdot h^3/64$
Parabolic section: $y_0 = 3h/5$ $\quad x_0 = 3b/8$	$A = 2b \cdot h/3$	$I_0 = \dfrac{8}{175} b \cdot h^3$

each small elemental surface components with their respective distances from the coordinate line. Generally this can be performed only by methods of integral calculus. However, for practical work it is sufficient to refer to tables showing values of I_0 for simple areas. Table 3-1 contains such information. For coordinate lines

other than those crossing the centroid second moments can be computed by

$$I_c = I_0 + Ac^2 \qquad (3.4)$$

where I_c is the moment of inertia with respect to the coordinate line, I_0 is the moment of inertia with respect to the parallel coordinate line crossing the

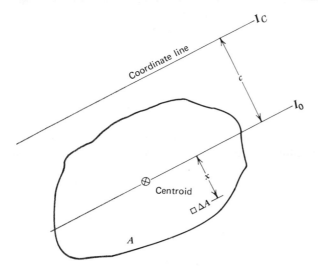

FIGURE 3.4   Interpretation of the moment of inertia. $I_0 = \Sigma\, x(\Delta A)$. $I_c = I_0 + Ac^2$.

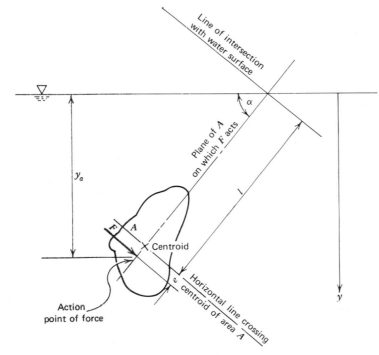

FIGURE 3.5   Action point of hydrostatic force.

centroid, A is the area, and c is the perpendicular distance between the two coordinate lines. Figure 3.4 depicts these variables.

Once the concept of first and second moments of areas are known we are ready to determine the location of the point where the resultant hydrostatic force acts on any known plane surface. If we refer to Figure 3.5 for our notations, the distance e between centroid and the action point of the hydrostatic force is given by

$$e = \frac{I_0}{lA} \tag{3.5}$$

where I_0 is the second moment of area A with respect to the centroid, l is the distance between centroid and the line of intersection of the plane of A with the water level, and e is the distance between the centroid and the point of action of the hydrostatic force measured on the surface on which it acts.

EXAMPLE 3.2

A rectangular gate 4 by 4 ft in size is set on a 45° plane with respect to the water surface; the centroid of the gate is 10 ft below the water surface (Figure E3.2). Compute the required location of

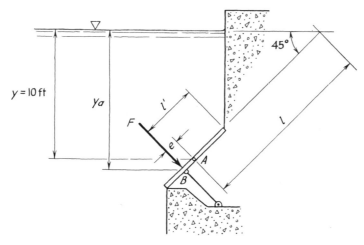

FIGURE E3.2

the horizontal axis at which the gate is to be hinged if the hydrostatic pressure acting on the gate is to be balanced around the hinge so that there will be no moment causing rotation of the gate under the loading specified.

SOLUTION:

Since the centroid of the gate is 10 ft under the water surface so that point A in the plane of the gate is at a distance l from the water level. Then

$$l = \frac{10}{\sin 45°}$$

$$= \frac{10}{0.707} = 14.14 \text{ ft}$$

Under static conditions the location of hinge has to be at the point where the resultant force acts on the gate. From Equation 3.5 the distance of the application of the hydrostatic pressure is e in which

$$e = \frac{I_0}{lA}$$

Where

$$I_0 = b \cdot \frac{h^3}{12} \qquad \text{(from Table 3-1)}$$

$$= \frac{4(4)^3}{12} = 21.33 \text{ ft}^4$$

$$A = b \cdot h = 4 \times 4 = 16 \text{ ft}^2$$

Hence

$$e = \frac{21.33}{(14.14)(16)} = 0.0943 \text{ ft} = 1.13 \text{ in.}$$

This is the distance from the centroid of the gate measured in the same plane. Hence point B is located at distance

$$l' = (2 \times 12) + 1.13 = 25.13 \text{ in.}$$

measured from the top edge of the gate.

The magnitude of the hydrostatic force F acting on a plane surface of arbitrary orientation is the product of the hydrostatic pressure acting at the centroid of the plane area and the magnitude of the area, or

$$F = A \cdot p_c = A \cdot \gamma \cdot y_c \qquad (3.6)$$

where y_c is the depth of water above the centroid.

The location where this force acts is in the same vertical plane as the centroid. The depth of water over the action point is given by

$$y_a = (1 + e) \sin \alpha \qquad (3.7)$$

For a vertical plane α equals 90° in which case $(l + e)$ equals y_a shown in Figure 3.5. In Table 3-2 the location of the action point below the centroid is given for some vertical geometric arrangements often found in practice.

Table 3-2 Location of Action Point of Hydrostatic Force on Vertical Gates

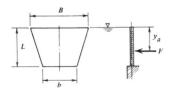

$$y_a = \frac{2B + 11b}{B + 2b} \cdot \frac{L}{6}$$

$$y_a = \tfrac{2}{3}L$$

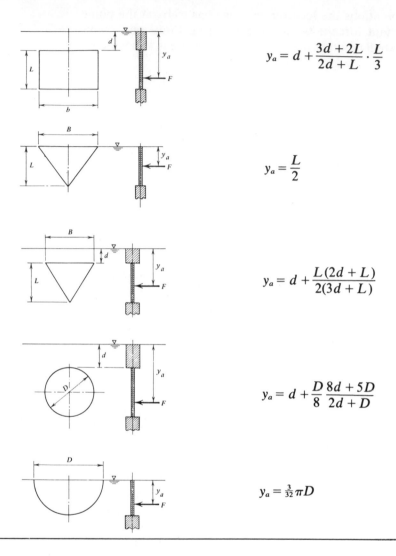

$$y_a = d + \frac{3d + 2L}{2d + L} \cdot \frac{L}{3}$$

$$y_a = \frac{L}{2}$$

$$y_a = d + \frac{L(2d + L)}{2(3d + L)}$$

$$y_a = d + \frac{D}{8}\frac{8d + 5D}{2d + D}$$

$$y_a = \tfrac{3}{32}\pi D$$

EXAMPLE 3.3

Calculate the total hydrostatic force acting on the gate's two hinges in the previous example problem.

SOLUTION:

From Equation 3.7

$$F = \gamma A l \sin \alpha$$
$$= 62.4(4 \times 4)(14.14) \sin 45°$$
$$= 10{,}000 \, lb$$

This is the total hydrostatic force acting on the gate. Each hinge carries one-half of this load; hence they are to be designed for a load of 5000 lb.

EXAMPLE 3.4

A 1 m diameter flood gate placed vertically is 4 m below the water level at its highest point. The gate is hinged at the top. Compute the required horizontal force to be applied at the bottom of the gate in order to open it. Assume that the pressure on the other side of the gate is atmospheric pressure and neglect the weight of the gate.

SOLUTION:

We know that the

$$\gamma \text{ of water} = 1000 \text{ kg/m}^3$$

$$\text{Area of the gate} = \frac{\pi}{4}(1)^2$$

$$A = 0.7855 \text{ m}^2$$

The depth of the centroid of the gate is $l = 4.5$ m. Next determine the location of F_1 acting on the gate.

From Equation 3.5

$$e = \frac{I_0}{lA}$$

From Table 3-1

$$I_0 = \frac{\pi R^4}{4}$$

$$= \frac{\pi (0.5)^4}{4} = 0.0491 \text{ m}^4$$

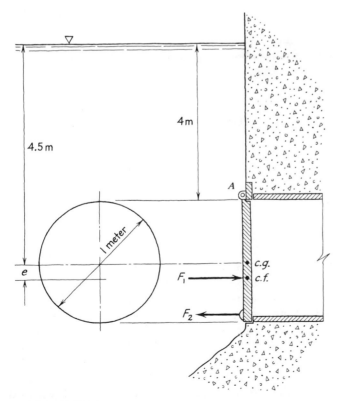

FIGURE E3.4

Therefore,

$$e = \frac{0.0491}{4.5(0.7855)} = 0.0139 \text{ m}$$

The force F_1 acting on the gate is located at a depth $y_a = (l + e)$. This equals $4.5 + 0.0139$, which equals

$$y_a = 4.5139 \text{ m}$$

The magnitude of this hydrostatic force is

$$F_1 = p_c A = 4.5 \times 1000 \times 0.7855 = 3534.75 \text{ kg}$$

To determine F_2 we take the moment of the two forces with respect to the hinge at A:

$$F_2 \times D = F_1 \left(\frac{D}{2} + e \right)$$

$$F_2 = \frac{3534.75(0.5 + 0.0139)}{1}$$

$$= 1816.51 \text{ kg}$$

CURVED SURFACES

Hydrostatic forces acting on curved surfaces are somewhat more difficult to evaluate. In such cases a graphical approach to the computations is advantageous. Consider, for example, the segment gate shown in Figure 3.6a. As the normal lines of all surface elements cross at the pivot point of the gate—the center of the circle of which the gate is a segment—it is obvious that the resultant force will go through this pivot point also. Also note, however, that at each surface point the magnitude of the pressure as well as its direction varies. As such, it is convenient to separate the horizontal and vertical components of the pressure. This results as shown in Figure 3.6b, in two rather simple pressure diagrams: a triangular one for the horizontal pressures and a segment-shaped one for vertical pressures. For the latter, depths below the water level at various points on the gate need to be multiplied by the unit weight of the water as a scaling factor in order to express vertical pressures. To find the horizontal and vertical components of the resultant hydrostatic force, we only need to find the points of action of the two pressure diagrams through which the resultant must pass. By the laws of statics these component forces intersect at a point through which their resultant force passes. To find this point by graphical means is a simple matter, as is the graphical addition of these two components in order to obtain the resultant.

BUOYANCY

For floating objects the concepts developed in the previous paragraph apply also. As shown in Figure 3.7, horizontal force components on the two sides of the body cancel each other. The vertical components all but cancel each other except for the volume where the floating body displaces water, resulting in a force acting upward. Accordingly, this buoyant hydrostatic force on a floating body of any shape equals the volume of the body multiplied by the unit weight of water acting at the displaced mass center. This is the famous law of Archimedes. Of course, the floating body itself will also have its own weight, a force acting downward opposing the lifting force of the water. Whether a body will sink to the bottom, rise to the top, or experience an apparent weightlessness in water depends on whether the hydrostatic uplift is smaller, larger, or equal to the body's weight. This is the principle behind the operation of submarines.

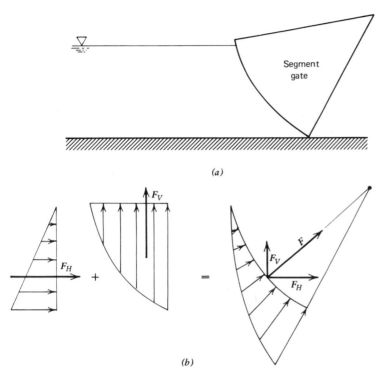

FIGURE 3.6 Hydrostatic pressure on curved surfaces.

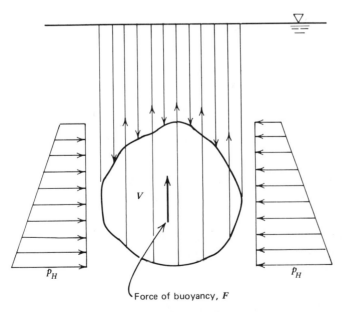

FIGURE 3.7 The force of buoyancy. $F = \gamma \forall$. $\forall$ = volume of body.

EXAMPLE 3.5

A 100,000 deadweight ton sunken ship is to be raised from 200 ft to the water surface by pumping air into its sealed holds. How much water is to be expelled from the ship to make it float?

SOLUTION:

We know that 10^5 tons equal 2×10^8 lb, and that 1 ft³ of water weighs 62.4 lb. Hence,

$$2 \times 10^8 \text{ lb of water}$$

equals

$$3.205 \times 10^6 \quad \text{ft}^3$$

which is the minimum amount of air space required.

The required air pressure to expel the water from the ship's holds

$$200 \text{ ft} \times 62.4 \text{ lb/ft}^3 = 12,480 \text{ lb/ft}^2$$
$$= 60,940 \text{ psi}$$

Bodies floating on the water surface will submerge to a point at which the weight of the displaced water will equal their own weight. In order that such a floating object be in a stable condition, it is required that the mass center of the floating object be below the mass center of the displaced water. If the two mass centers coincide, the object will be unstable. Overloaded boats capsize because their mass center, including that of the load, is higher than that of the displaced water.

PROBLEMS

3.1   A bridge pier 50 by 12 ft is built such that its base is 12 ft below water level. Determine the hydrostatic uplift force.

3.2   On a closed vertical sluice gate that is 8 ft wide and 10 ft deep below water level hydrostatic force acts from the upstream side only. Determine the point of action and the magnitude and direction of the force.

3.3   On the gate described in Problem 3.2 determine the combined action of both upstream and downstream hydrostatic forces if the water depth at the downstream equals 5 ft.

3.4   A circular flood gate, 2.5 ft in diameter is under 10 ft of water with respect to its center. Determine the pressure distribution, the resultant force, and the location of the action point.

3.5   A rectangular 4 ft square gate is on a 45° slope with its side parallel to the water surface. The center of the gate is under 18 ft of water. Compute the magnitude and the point of attack of the hydrostatic force.

3.6   Water is carried in a closed tank of the shape of a cube, 6 ft on each side. Determine the maximum hydrostatic pressure acting on the bottom if the truck carrying the tank stops with a deceleration of $-g$.

3.7   Determine the first and second moments of a rectangular area 3 m on each side with respect to an axis parallel to one side and 6 m away from the centroid.

3.8   A 25 ft diameter tunnel is built in sand. The ground water level in the sand is 35 ft above the top of the tunnel. Determine the maximum force due to the pressure of the water and its location in the tunnel wall.

3.9   The floor of a watertight basement of a 30 by 60 ft house is 6 ft below the ground water level. Assuming that the density of the concrete floor is 2.6 grams/cm^3 determine the required floor thickness to prevent uplift with a factor of safety of 1.5.

3.10   A 12 ft^2 floating dock has a 2000 lb dead weight. How many 3.5 ft tall 50 gallon steel drums should be placed under the platform in order that they may sink to their center line into the water, if the weight of one steel drum is 75 lb.

FLOW IN PIPES

The great majority of industrial and municipal applications of hydraulics involve flow in pipes. This chapter first reviews the various methods used in the computation of energy losses in pipes. Then the design of actual piping systems is explained. Water hammer and other unsteady flow problems are also discussed.

To supply water, pipelines were built sporadically from the Middle Ages on. The water was led from springs in the mountains to nearby forts and palaces, by gravity flow. With the development of iron and steel manufacturing and after the introduction of efficient pumps more than 150 years ago, the building of pipelines became common practice. The amount of energy required to push the desired amount of discharge through a pipeline became a paramount question for designers. Because the discharge as well as the pressure lost through a pipe were easily measurable quantities, a great deal of experimental and field data was collected on pipe flow. A natural scientific development to put this data into the form of design formulas followed quite early.

THE EMPIRICAL METHOD

Plotting the data on graphs and developing empirical equations was one of the ways to simplify design. At least one of these early equations is still widely used in practice. This is the so-called Hazen-Williams equation, which may be written in a dimensionless form, valid for any consistent units of measurements, as

$$\frac{V}{\sqrt{gR}} = 0.232CR^{0.13}S^{0.54} \quad (4.1a)$$

in which V is the average velocity in the pipe, S is the slope of the energy grade line—that is, the loss of energy per unit length of the pipe—R is the so-called hydraulic radius (Equation 8.17) which, for circular pipes, is equal to one-fourth of the pipe diameter. The coefficient C is a friction factor with the actual dimension of length on the minus 0.13th power, introduced as a constant to represent the smoothness of the pipe walls. Table 4-1 provides a list of C coefficients for various types of pipe walls. Generally the magnitude of the coefficient is taken as 100 for average conditions, while its value could be as low as 50 for badly corroded

Table 4-1 Values of C for the Hazen-Williams Formula

Type of Pipe	C
Asbestos cement	140
Brass	130–140
Brick sewer	100
Cast iron	
New, unlined	130
Old, unlined	40–120
Cement lined	130–150
Bitumastic enamel lined	140–150
Tar-coated	115–135
Concrete or concrete lined	
Steel forms	140
Wooden forms	120
Centrifugally spun	135
Copper	130–140
Fire hose (rubber lined)	135
Galvanized iron	120
Glass	140
Lead	130–140
Plastic	140–150
Steel	
Coal-tar enamel lined	145–150
New unlined	140–150
Riveted	110
Tin	130
Vitrified clay	100–140

pipes, and as high as 150 for very smooth plastic or glass pipes.

In the English system, in which the Hazen-Williams formula is most often used, Equation 4.1a is written as

$$Q = 0.285 CD^{2.63} S^{0.54} \qquad (4.1b)$$

in which Q is to be used in gallons per minute, and D in inches. Figure 4.1 is a graphical representation of Equation 4.1b to facilitate computations.

Note that the Hazen-Williams formula may be used for water only, as it does not contain any terms related to the physical properties of the fluid. Although the formula is incorrect from theoretical viewpoints, it still seems to give acceptable results in practice. The reason for this is the built-in uncertainty in the determination of the C coefficient. This uncertainty allows for an apparent but faulty credence.

EXAMPLE 4.1

A 3000 ft long pipe of 10 in. diameter carries 5 cfs of water. The C coefficient is assumed to be 100. Calculate the loss of energy due to friction using the Hazen-Williams nomograph shown in Figure 4.1.

SOLUTION:

The variables are:

$$Q = 5 \text{ cfs} = 300 \text{ ft}^3/\text{min}$$
$$D = 10 \text{ in.}$$
$$C = 100$$

In the nomograph connect Q and D with a straight line and mark off the intersection on the turning line. Connect the marked point with C and extend the line to S. The resulting S is 1/1000 ft per ft. For 3000 ft the head loss is $3000 \times 0.001 = 3$ ft.

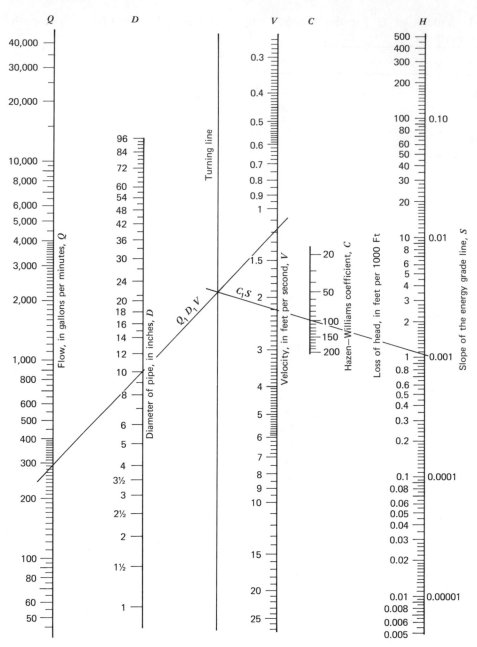

FIGURE 4.1 The Hazen-Williams nomograph. (Courtesy of Public Works Magazine.)

THE SCIENTIFIC METHOD

Scientific speculations concerning the physical relationships controlling pipe flow date back to the middle of the nineteenth century. Fundamental concepts were laid down by Chézy whose

formula

$$V = \frac{4Q}{\pi D^2} = C\frac{1}{2}\sqrt{DS} = C\sqrt{RS} \quad (4.2a)$$

is also used for open channel flow. Equation 4.2a states that the discharge Q of a pipe is proportional to the cross-sectional area A of the pipe, and to a friction factor C, which is again a coefficient of uncertainty. The terms under the radical sign are the pipe diameter D and the slope of the energy line S. S is the ratio of the energy loss h_L to the length L of the pipe. Rearranging Equation 4.2a we may write

$$S = \frac{h_L}{L} = \left(\frac{8Q}{C\pi}\right)^2 D^{-5}$$

from which

$$h_L = \left(\frac{8}{C\pi}\right)^2 \frac{L}{D^5} Q^2 \qquad (4.2b)$$

This means that the head loss in a pipe is directly proportional to the square of the discharge, and inversely proportional to the pipe diameter raised to the fifth power, a very influential factor. The practical value of Equation 4.2 is still limited by the unknown C constant, which, as it was soon proved, was not a constant at all.

More satisfactory from a conceptual standpoint was the approach taken by Darcy and Weisbach. Their formula

$$h_L = f\frac{L}{D}\frac{V^2}{2g} \qquad (4.3)$$

states that the head loss is proportional to the kinetic energy, $V^2/2g$ of the flow and to the length of the pipe and inversely proportional to the diameter. The friction factor f was subsequently subject to much theoretical and experimental research.

The value of f was found to be dependent on the Reynolds number of the flow and the relative roughness of the pipe. The Reynolds number is a dimensionless term that expresses the ratio of the inertial energy that drives the fluid to the viscous energy loss that resists the movement. This number is computed by the formula

$$\mathbf{R} = \frac{\rho D V}{\mu} \qquad (4.4)$$

where ρ is the density and μ is the absolute viscosity of the fluid.

For flows of very small velocity, for fluids of high viscosity, or for very small pipe diameters, the Reynolds number may be less than 2300. In this case the flow is smooth and ordered, the pathlines in the fluid are parallel, and dyes injected into the fluid travel along their respective flow lines without mixing into the whole fluid. Flows of this type are called laminar. The friction factor in laminar flow is computed from

$$f = \frac{64}{\mathbf{R}} \qquad (4.5)$$

showing a linear relationship between the Reynolds number and the friction factor.

In flows at higher Reynolds numbers turbulence sets in. When the flow is turbulent the fluid particles move about in the pipe in a random manner, completely mixing up with the dye that may be injected into the flow. The higher the Reynolds number the higher the rate of turbulence. This random mixing of fluid particles consume a considerable amount of energy in the flow. Hence, proportionally more energy is needed to move turbulent flow than a laminar one. In turbulent flow the influence of the roughness of the pipe wall becomes dominant. The roughness size e is expressed in terms of the pipe diameter as

$$\epsilon = \frac{e}{D} \qquad (4.6)$$

which is called relative roughness. The actual measurement of e is obviously not possible; rather, the value is computed backward from experimental measurements.

The first extensive determination of the effect of pipe roughness upon the flow was done by Nikuradse. He has coated the inside of pipes of different diameters with sands of various grain sizes. Based on his experiments the term "equivalent sand roughness" came into use. Table 4-2 lists the equivalent sand roughness values for various commercially available pipe materials. Results of Nikuradse's experiments indicated that when the roughness of the pipe fully influences the flow at high

Table 4-2 Equivalent Sand Roughness for Pipe Materials

Commercial Pipe Surface, New	Equivalent Sand Grain Roughness, e in ft	e^3 in ft^3
Glass, drawn brass, copper, lead	Smooth	—
Wrought iron, steel	1.5×10^{-4}	3.4×10^{-12}
Asphalted cast iron	4.0×10^{-4}	64×10^{-12}
Galvanized iron	5.0×10^{-4}	124×10^{-12}
Cast iron	8.5×10^{-4}	614×10^{-12}
Concrete	10–100×10^{-4}	10^{-9}–10^{-6}
Riveted steel	30–300×10^{-4}	10^{-8}–10^{-5}

Reynolds numbers, the friction factor is independent of the Reynolds number. Von Kármán used the theory of boundary layer to derive a formula for turbulent flow within rough pipe boundaries. The resulting equation for the value of the friction factor is

$$\frac{1}{\sqrt{f}} = 1.14 + 2 \log \frac{D}{e} \qquad (4.7)$$

The constant in this equation was adjusted from its theoretical value to correspond to the experimental results of Nikuradse.

Other investigators developed additional formulas for expressing Nikuradse's experimental results in mathematical forms. Moody has plotted these equations on a graph similar to the one reproduced in Figure 4.2. It must be

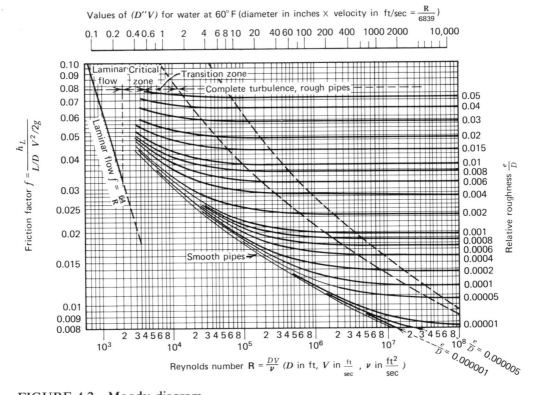

FIGURE 4.2   Moody diagram.

noted that both experimental and theoretical investigations led to uncertain results in the transitional range when the flow changes from laminar to turbulent conditions. From a practical point of view this is rather immaterial because almost all designs concern flow well in the turbulent range.

With the Darcy-Weisbach equation (Equation 4.3) and with Moody's diagram (Figure 4.2) available, frictional losses in pipes can be computed with relative ease and reasonable accuracy. Computations

using these tools may seek to determine either the energy loss h_L, the discharge Q, or the diameter D required. In all of these three cases all other pertinent variables are presumed to be known.

When the energy loss h_L is sought for known discharge and known pipe, (Q, D, e, and L are known) the Reynolds number and the relative roughness are readily determined by substituting into Equations 4.4 and 4.6. These will allow us to read off the value of the friction factor f directly from Figure 4.2.

EXAMPLE 4.2

A 12 km long pipe carries 0.2 m³/s of water. The pipe diameter is 30 cm; its relative roughness is 0.004. Compute the change in head loss if the temperature of the water changes from 65°C to 30°C.

SOLUTION:

The average velocity of the flow $= \dfrac{Q}{A} = \dfrac{0.2}{\dfrac{\pi}{4}(0.3)^2} = 2.829 \dfrac{m}{s}$

The equation for head loss $h_L = f\dfrac{L}{D}\dfrac{V^2}{2g}$

From Table 1-5, at 30°C

$$\mu \cong 0.8807 \text{ centipoise}$$
$$= 0.8807 \times 10^{-3} \text{ poise}$$
$$= 8.807 \times 10^{-5} \frac{N.s}{m^2}$$

At 65°C

$$\mu \cong (0.4276 \times 10^{-3}) \text{ poise}$$
$$= 4.284 \times 10^{-5} \frac{N.s}{m^2}$$

From Table 1-2, at 30°C

$$\rho \approx 0.9970 \frac{g}{cm^3} = 997 \frac{kg}{m^3}$$
$$= \frac{997}{9.81} = 101.63 \frac{N.s^2}{m^4}$$

At 65°C

$$\rho \cong 0.980 \text{ g/cm}^3 = 980 \text{ kg/m}^3$$
$$= \frac{980}{9.81} = 99.897 \frac{N.s^2}{m^4}$$

From Equation 4.4

$$R = \frac{\rho VD}{\mu}$$

Then

$$R \text{ at } 30°C = \frac{101.63 \times 2.829 \times 0.3}{8.807 \times 10^{-5}} = 9.794 \times 10^5$$

$$R \text{ at } 65°C = \frac{99.897 \times 2.829 \times 0.3}{4.284 \times 10^{-5}} = 19.791 \times 10^5$$

The relative roughness $e/D = 0.004$.
Entering Moody's diagram we find that because of the large Reynolds numbers

$$f \text{ at } 65°C \cong f \text{ at } 30°C = 0.0282$$

The head loss may be computed from Equation 4.3

$$h_L = f\frac{L}{D}\frac{V^2}{2g} = 0.0282\frac{(12,000)}{0.3}\frac{(2.829)^2}{2 \times 9.81}$$
$$= 460.0 \text{ m}$$

over the entire length of the 12 km pipeline.

If the discharge Q is the unknown, the computation is not as direct as the one above. The problem is that with Q unknown both the velocity V and the friction factor f are also unknown. Hence we have three unknowns. Known values are e, D, and L, from which the relative roughness e/D can be computed. Having this value we determine a trial value for f by inspecting the Moody diagram. When this f is substituted into the Darcy-Weisbach equation, a velocity V can be computed which, of course, is still a trial value. This trial V can now be put into the Reynolds number; this allows us to find an improved value for f from the Moody diagram. If this new friction factor closely corresponds to its first trial value (say within two significant figures) V may be assumed correct. If there is a significant difference between the two f values, a new velocity should be computed using the latter f value. If after this second trial the friction factor appears to be reasonably correct, the velocity V is also correct. This allows us to determine the discharge Q by multiplying V with the area of the pipe.

EXAMPLE 4.3

An 8 in. cast-iron pipe carries water at 40°F from reservoir A and discharges it into reservoir B (Figure E4.3). The length of the pipe is 5000 ft. The elevation of water surface of A is 3300 and 3100 ft for B. Calculate the discharge of the pipe.

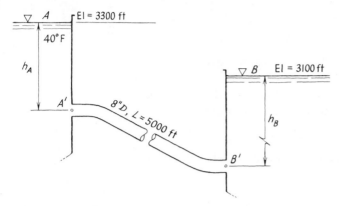

FIGURE E4.3

SOLUTION:

From Table 4-2 the roughness of cast-iron pipe $e = 0.00085$ ft. Then, from Equation 4.6

$$\epsilon = \frac{e}{D} = \frac{0.00085}{8/12}$$

$$= 0.00127$$

From the properties of water at 40°F (Table 1-5)

$$\mu = 3.229 \times 10^{-5} \frac{\text{lbf sec}}{\text{ft}^2}$$

From Equation 1.2

$$\rho = 1.94 \frac{\text{slug}}{\text{ft}^3}$$

Then

$$\nu = 1.664 \times 10^{-5} \frac{\text{ft}^2}{\text{sec}}$$

The Reynolds number is computed as

$$\mathbf{R} = \frac{VD}{\nu} = 4.006 \times 10^4 V$$

The head loss in the pipe is $3300 - 3100 = 200$ ft over the 5000 ft pipe length. Hence the Darcy-Weisbach formula will take the form

$$h_L = f \frac{L}{D} \frac{V^2}{2g}$$

$$200 = f \frac{5000}{0.666} \frac{V^2}{64.4}$$

Assume a value for f, say, $f = 0.03$; substituting this into the above formula we obtain

$$V^2 = 85.85$$

and

$$V = 9.26 \text{ fps}$$

Substituting v into $\mathbf{R}$, we obtain

$$\mathbf{R} = 3.712 \times 10^5$$

With this $\mathbf{R}$ value and with $e/D = 0.00127$ we enter Moody's diagram and find that the corresponding f equals 0.0215, which is not equal to the assumed f value. Substituting the new f into the Darcy-Weisbach equation we get $v = 8.93$ fps. Hence the new $\mathbf{R}$ value is

$$\mathbf{R} = 3.57 \times 10^5$$

With this new $\mathbf{R}$, Moody's diagram results in an f of 0.0213, which is a sufficiently close result.

The discharge is then equal to

$$Q = V \cdot A = 8.93 \, \frac{\pi}{4} \left(\frac{8}{12}\right)^2 = 3.117 \text{ cfs}$$

When the pipe diameter D is sought the computation is somewhat more involved. In this case neither the relative roughness, nor the Reynolds number, and not even the velocity can be computed directly. With Moody's diagram a tedious trial and error method is used to determine D. But a direct solution for D is possible by using the modified Moody diagram designed by Wen-Hsiung Li, shown in Figure 4.3a. In this graph the variables are arranged so that D appears only once. To use the graph first, the term

$$\left(\frac{gS}{Q^2}\right)^{1/5} \tag{4.8}$$

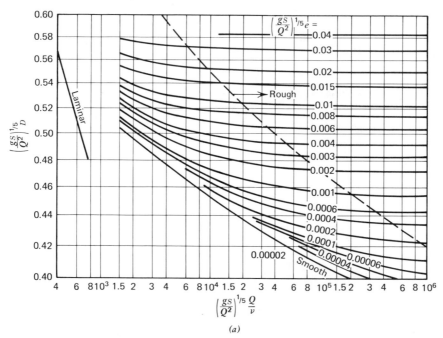

FIGURE 4.3 (a) Li's pipe flow diagram. (From "Direct Determination of Pipe Size by Wen-Hsiung Li, *Civil Engineering*, American Society of Civil Engineers, Vol. 44, No. 6 June 1974.)

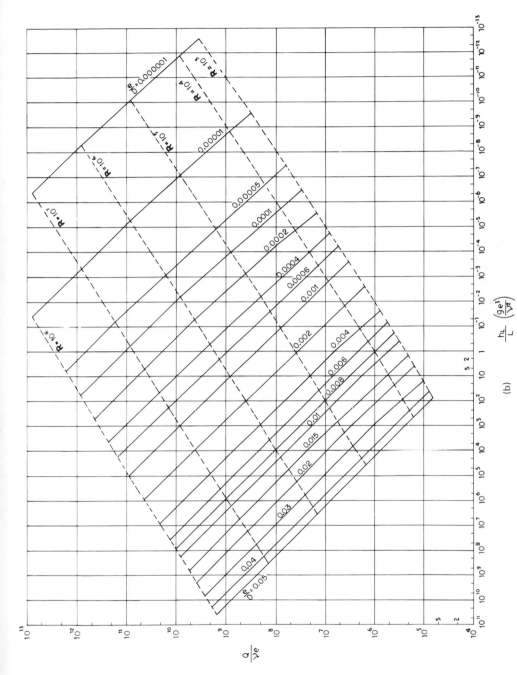

FIGURE 4.3 (*Cont'd*) (*b*) Asthana's pipe flow diagram. (Courtesy of K. C. Asthana, University of Addis Ababa, Addis Ababa, Ethiopia.)

53

must be computed. All coordinates of the graph in Figure 4.3a are expressed as functions of this term. By substituting the known values of Q, v, and e, the diameter D is simply obtained from the graph.

By combining the Darcy-Weisbach equation and the Moody diagram (Equation 4.3 and Figure 4.2) into a single graph K.

C. Asthana introduced a further simplification. His graph, shown in Figure 4.3b, plots the Reynolds number and the relative roughness in the coordinates of $(h_L/L)(ge^3/v^2)$ and Q/ve. Values of e and e^3 are shown in Table 4-2 for various pipe materials.

EXAMPLE 4.4

Water moving at 900 gpm and at 65°F is to be transferred over 2000 ft to an elevation 60 ft lower than the source of supply (Figure E4.4). Determine the required diameter of the concrete pipe such that the friction loss is equal to the difference in elevation.

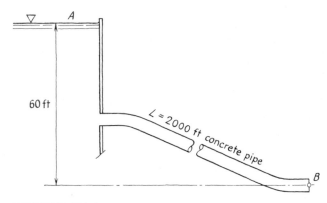

FIGURE E4.4

SOLUTION:

From the properties of water (Chapter 1), at 60°F

$$\rho = 1.938 \text{ slug/ft}^3, \qquad \mu = 0.00002359 \text{ lbf sec/ft}^2,$$

and

$$v = 1.217 \times 10^{-5} \text{ ft}^2/\text{sec}.$$

The slope is

$$\frac{h_L}{L} = S = \frac{60}{2000} = 0.03$$

From Table 4-2 the average concrete roughness is $e = 5 \times 10^{-3}$ ft.

$$Q = 900 \text{ gpm} \times 2.23 \times 10^{-3} = 2.00 \text{ cfs}$$

We may use Li's graph and Equation 4.8

$$\left(\frac{gS}{Q^2}\right)^{1/5} = \left(\frac{32.2 \times 0.03}{2.0^2}\right)^{1/5} = 0.2398^{0.2}$$

$$= 0.7515$$

From Figure 4.3a we see that the following products are needed to use the graph:

$$\left(\frac{gS}{Q^2}\right)^{1/5}\frac{Q}{\nu} = 0.7515\frac{2.0}{1.217\times 10^{-5}} = 1.239\times 10^{5}$$

$$\left(\frac{gS}{Q^2}\right)^{1/5}e = 0.7515(5\times 10^{-3}) = 3.75\times 10^{-3}$$

Entering Figure 4.3a we find

$$\left(\frac{gS}{Q^2}\right)^{1/5}D = 0.485$$

Hence the required diameter is

$$D = \frac{0.485}{0.7515} = 0.645 \text{ ft, say, 8 in.}$$

We may solve the same problem by using Asthana's graph shown in Figure 4.3b. For this we first obtain

$$\frac{h_L}{L}\left(\frac{ge^3}{\nu^2}\right) = 0.03(32.2)\frac{(5\times 10^{-3})^3}{(1.217\times 10^{-5})^2}$$

$$= 8.1\times 10^{2}$$

and

$$\frac{Q}{\nu e} = \frac{2.0}{(1.217\times 10^{-5})5\times 10^{-3}}$$

$$= 3.3\times 10^{7}$$

Entering Figure 4.3b with these coordinates we find

$$e/D = 0.0078 \qquad \text{and} \qquad \mathbf{R} = 2\times 10^{6}$$

Since e is known, it follows that

$$D = \frac{5\times 10^{-3}}{7.8\times 10^{-3}} = 0.64 \text{ ft, say, 8 in.}$$

The use of the Darcy-Weisbach equation along with Moody's diagram can provide the best, most reliable solutions for pipe flow problems. From the standpoint of practical work two criticisms may be raised. First, the roughness of the pipe is still an unknown, particularly if we consider the long time in which a pipeline may be in use. This results somewhat in the same uncertainty as the less scientific but direct computation using the old Hazen-Williams nomograph. The other criticism may be raised because of the cumbersome trial and error procedure just mentioned. Because of this difficulty the use of the Moody diagram is very limited in practice.

Recent surveys indicate that as much as 80% of engineers in the water supply practice prefer to use old nomographs instead of the Darcy-Weisbach equation.

THE CONVEYANCE METHOD

The scientific advantages of the Darcy-Weisbach approach and the precision at which the f friction factor was determined make the scientific method a valuable one. If we recognize that in most technical applications pipelines are designed for a possible maximum design discharge and that such maximum discharge will surely be carried in the pipe at fully turbulent condition, the shortcomings may be avoided by the following development. Taking von Kármán's formula for f for turbulent flow, Equation 4.7 and substituting it into Equation 4.3, we obtain

$$Q = \left[\frac{\pi}{4} \sqrt{2g} \left(2 \log \frac{D}{e} + 1.14 \right) D^{2.5} \right] \left(\frac{h_L}{L} \right)^{1/2} \quad (4.9)$$

$$Q = K \left(\frac{h_L}{L} \right)^{1/2} = K \sqrt{S} \quad (4.10)$$

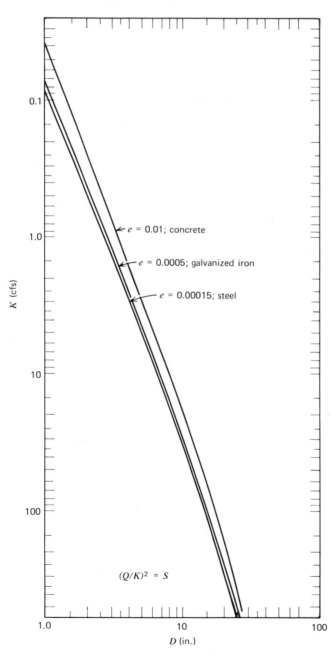

FIGURE 4.4   Pipe conveyance graph.

in which all terms in the brackets are known for a particular pipe. As such, the terms in the brackets may be lumped to form a new parameter called conveyance, K, as shown in Figure 4.4. Equation 4.10 allows us to solve quickly and simply most pipe flow problems, without the sometimes cumbersome procedure associated with the use of Moody's diagram.

Continuous, uniform distribution of water along the pipes of a municipal water supply system or in a network of sprinkler irrigation pipes may be considered by the conveyance method. Let us consider a pipe of K conveyance and L length along which a discharge Q_u is distributed uniformly such that

$$q = \frac{Q_u}{L} \qquad (4.11)$$

where q is the discharge delivered in a unit length of pipe. The continuity equation for such pipe takes the form of

$$Q_{in} = Q_{out} + Q_u \qquad (4.12)$$

As the discharge decreases along the pipe the velocity decreases also. Hence the rate of energy loss is not a constant for the distributing pipe but a decreasing value resulting in a curved energy grade line. The two end points of this curved energy grade line represent the inflow and outflow energy levels denoted by H_{in} and H_{out}, respectively. The value of the conveyance coefficient, however, is a function of the pipe roughness and pipe diameter only; therefore, K is constant for the pipe regardless of the discharge flowing.

This allows us to write Equation 4.10 for a point x along the pipe in the form of

$$S_{(x)} = \frac{h_L}{x} = \left(\frac{Q}{K}\right)^2 = \frac{(Q_{in} - (x/L)Q_u)^2}{K^2}$$

Expanding the squared term in the numerator of the right side and solving the equation by integration between $x = 0$ and $x = L$, and then substituting H_{in} and H_{out} for the limits of h, we obtain

$$h_L = H_{in} - H_{out}$$
$$= \frac{L}{K^2}\left(Q_{in}^2 - Q_{in} \cdot Q_u + \frac{1}{3}Q_u^2\right) \quad (4.13)$$

which is a general solution for a continuously distributing pipe. A special case is when the discharge is completely distributed along the line, for which Q_{out} equals zero and Q_{in} consequently equals Q_u. For this case Equation 4.13 takes the form of

$$h_L = H_{in} - H_{out} = \frac{1}{3}\left(\frac{L}{K^2}\right)Q_u^2 \quad (4.14)$$

By observing this equation we note that for continuous and complete discharge distribution the head loss in the pipeline is only one-third of the loss of the corresponding conventional pipe flow.

The advantage of the conveyance method is obvious from both a theoretical or a practical standpoint. It is based on the best theoretical results and is easy to use. Its limitations correspond to those of Equation 4.7; hence, it will not give reliable results for flows within the transitional region. But neither will Moody's diagram.

EXAMPLE 4.5

Calculate the conveyance of a 3 ft diameter concrete pipe.

SOLUTION:

From Equation 4.10

$$Q = K\sqrt{S}$$

where

$$S = \frac{h_L}{L}$$

$$K = \text{conveyance}$$

$$= \left[\frac{\pi}{4} \sqrt{2g} \left(2 \log \frac{D}{e} + 1.14 \right) D^{2.5} \right]$$

For the concrete pipe 3 ft in diameter, $e = 0.004$ ft from Table 4.2. Then

$$K = \left[\frac{\pi}{4} \sqrt{2g} \left(2 \log \frac{3}{0.004} + 1.14 \right) 3^{2.5} \right]$$

$$= [6.304(2 \times 2.875 + 1.14)15.59]$$

$$= 677.16*$$

The conveyance of the 3 ft concrete pipe $= 677.16$ cfs.

EXAMPLE 4.6

Determine the discharge of a 12 in. galvanized pipe if the length of the pipe is 1200 ft and the head loss is 100 ft.

SOLUTION:

The above given conditions imply that

$$h_L = 100 \text{ ft}$$
$$D = 1 \text{ ft (galvanized pipe)}$$
$$L = 1200 \text{ ft}$$

From Equation 4.10

$$Q = K\sqrt{S}$$

$$\sqrt{S} = \sqrt{\frac{h_L}{L}} = \left(\frac{100}{1200} \right)^{1/2} = 0.289$$

From Table 4.1 e for the galvanized pipe $= 0.0005$. Then

$$K = \left[\frac{\pi}{4} \sqrt{2g} \left(2 \log \frac{1}{0.0005} + 1.14 \right)(1)^{2.5} \right]$$

$$= [6.304(6.602 + 1.14)] = 48.81*$$

Hence

$$Q = 48.81 \times 0.289 = 14.11 \text{ cfs}$$

EXAMPLE 4.7

By using the conveyance method, determine the required pipe diameter for the problem described in Example 4.4.

---

* This value may be obtained directly from Figure 4.4.

SOLUTION:

From Equation 4.10 $Q = K\sqrt{S}$

$$\sqrt{S} = \sqrt{\left(\frac{60}{2000}\right)} = 0.1732$$

$$Q = 900 \text{ gpm} = 2 \text{ cfs}$$

Then

$$K = \frac{2}{0.1732} = 11.55$$

By entering this value into Figure 4.4 we obtain $D = 8$ in. pipe.

LOCAL LOSSES

In the hydraulic design of pipelines the energy lost through friction along the pipe is dominant for pipes of 100 ft (30 m) or longer. For shorter pipe lengths the aggregate of local energy losses at elbows, valves, inlet devices, and the like may be equal or more than the frictional losses along the pipe. For this reason it was necessary to develop formulas to compute local energy losses. In particular, this matter became important in the suction pipe of pumps in order that cavitation in the pump be prevented by careful design (Chapter 5).

Local losses in piping fixtures were found to be proportional to the amount of kinetic energy entering the fixture. The configuration of the fixture determines the constant of proportionality. Accordingly, local loss in a pipe fixture is computed by

$$h_v = k\frac{v^2}{2g} \qquad (4.15)$$

in which k is the so-called local loss coefficient and v is the velocity in the pipe before the fixture, unless otherwise specified. Table 4-3 lists local loss coefficients for a variety of fixtures.

Table 4-3 Local Loss Coefficients

Use the equation $h_v = kv^2/2g$ unless otherwise indicated. Energy loss E_L equals h_v head loss in feet.

	Perpendicular square entrance: $$k = 0.50 \quad \text{if edge is sharp.}$$	
	Perpendicular rounded entrance:	

$R/d =$	0.05	0.1	0.2	0.3	0.4
$k =$	0.25	0.17	0.08	0.05	0.04

	Perpendicular reentrant entrance: $$k = 0.8$$	
	Additional loss due to skewed entrance: $$k = 0.505 + 0.303 \sin \alpha + 0.226 \sin^2 \alpha$$	

Suction pipe in sump with conical mouthpiece:

$$E_L = D + \frac{5.6Q}{\sqrt{2g}D^{1.5}} - \frac{v^2}{2g}$$

Without mouthpiece:

$$E_L = 0.53D + \frac{4Q}{\sqrt{2g}D^{1.5}} - \frac{v^2}{2g}$$

Width of sump shown: $3.5D$

(After I. Vágás)

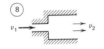

Strainer bucket:

$k = 10$ with foot valve

$k = 5.5$ without foot valve

(By Agroskin)

Standard Tee, entrance to minor line

$$k = 1.8$$

Sudden expansion:

$$E_L = \left(1 - \frac{v_2}{v_1}\right)^2 \frac{v_1^2}{2g}$$

or

$$E_L = \left(\frac{v_1}{v_2} - 1\right)^2 \frac{v_2^2}{2g}$$

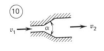

Sudden contraction:

$(d/D)^2 =$	0.01	0.1	0.2	0.4	0.6	0.8
k =	0.5	0.5	0.42	0.33	0.25	0.15

use v_2 in equation (4.15)

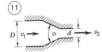

Confusor:

$$E_L = k(v_1^2 - v_2^2)/2g$$

$\alpha° =$	20	40	60	80
k =	0.20	0.28	0.32	0.35

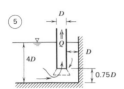

Diffusor:

$$E_L = k(v_1^2 - v_2^2)/2g$$

$\alpha° =$	6	10	20	40	60	80	100	120	140
k for $D = 3d$	0.12	0.16	0.39	0.80	1.0	1.06	1.04	1.04	1.04
$D = 1.5d$	0.12	0.16	0.39	0.96	1.22	1.16	1.10	1.06	1.04

(12) Sharp elbow:

$$k = 67.6 \times 10^{-6}(\alpha °)^{2.17}$$

(By Gibson)

(13) Bends:

$$k = (0.13 + 1.85(r/R)^{3.5})\sqrt{\alpha °/180°}$$

(By Hinds)

(14) Close return bend:

$$k = 2.2$$

(15) Gate valve:

$e/D =$	0	1/4	3/8	1/2	5/8	3/4	7/8
k =	0.15	0.26	0.81	2.06	5.52	17.0	97.8

(16) Globe valve:

$$k = 10 \quad \text{when fully open}$$
fully open

(17) Rotary valve:

$\alpha ° =$	5	10	20	30	40	50	60	70	80
k =	0.05	0.29	1.56	5.47	17.3	52.6	206	485	∞

(By Agroskin)

(18) Check valves:

Swing type	$k = 2.5$	When fully open
Ball type	$k = 70.0$	
Lift type	$k = 12.0$	

(19) Angle valve:

$$k = 5.0 \quad \text{if fully open}$$

(20) Segment gate in rectangular conduit:

$$k = 0.8 \; 1.3 \left[\left(\frac{1}{n}\right) - n\right]^2$$

where $n = \varphi/\varphi_0 =$ the rate of opening with respect to the central angle.

(By Abelyev)

Sluice gate in rectangular conduit:

$$k = 0.3\ 1.9\left[\left(\frac{1}{n}\right) - n\right]^2$$

where $n = h/H$.

(By Burkov)

Measuring nozzle:

$$E_L = 0.3\ \Delta p \qquad \text{for} \qquad d = 0.8D$$
$$E_L = 0.95\ \Delta p \qquad \text{for} \qquad d = 0.2D$$

where Δp is the measured pressure drop.

(By A.S.M.E.)

Venturi meter:

$$E_L = 0.1\ \Delta p \qquad \text{to} \qquad 0.2\ \Delta p$$

where Δp is the measured pressure drop.

Measuring orifice, square edged:

$$E_L = \Delta p \left(1 - \left(\frac{d}{D}\right)^2\right)$$

where Δp is the measured pressure drop.

Confusor outlet:

$d/D =$	0.5	0.6	0.8	0.9
$k\ =$	5.5	4	2.55	1.1

(By Mostkov)

Exit from pipe into reservoir:

$$k = 1.0$$

Diffusor outlet for $D/d > 2$:

$\alpha° =$	8	15	30	45
$k\ =$	0.05	0.18	0.5	0.6

(By Mostkov)

Equation 4.14 may be converted to express the energy loss through a valve or other fixture in terms of the discharge flowing through it. For this Equation 2.4 may be substituted as follows:

$$h_v = \left(\frac{8k}{g\pi^2 D^4}\right) Q^2 \qquad (4.16)$$

which could further be simplified and rearranged in the form of

$$Q = C_v \sqrt{h_v} \qquad (4.17)$$

where

$$C_v = \pi D^2 \sqrt{\frac{g}{8k}} \qquad (4.18)$$

in which C_v represents the conveyance of a hydraulic device creating a local loss in a piping system.

EXAMPLE 4.8

Using the Darcy-Weisbach equation, determine the discharge of a 4-in. cast-iron pipe of 150 ft length, if the system includes a strainer bucket with a foot valve, a gate valve half closed, a fully open globe valve, a square-edged measuring orifice reducing the diameter to 2 in. and a 2-in. diameter confusor at the outlet. The total head loss in the system is 25 ft. The measured head loss in the orifice may be taken as 0.2 ft.

SOLUTION:

For the given total head losses:

$h_T = 25$ ft

h_T = friction loss + loss due to strainer bucket with foot valve + gate valve half closed + globe valve fully open + square edge measuring orifice + confusor outlet

From Table 4-3, k values are determined as follows:

$k_1 = 10$ for strainer bucket with foot valve

$k_2 = 2.06$ for gate valve, half closed

$k_3 = 10$ for globe valve, fully open

$k_4 = 5.5$ for the confusor with $\dfrac{d}{D} = 0.5$

$h = 0.2$ ft for the square-edged measuring orifice loss.

For cast iron pipe $e = 0.00085$ ft.

Then

$$\epsilon = \frac{e}{D} = \frac{0.00085}{4/12} = 0.0026$$

Assume the water temperature is 60°F.

Then

$$\rho = 1.938 \, \frac{slug}{ft^3}$$

$$\mu = 2.359 \times 10^{-5} \, \frac{lb \; sec}{ft^2}$$

or

$$\nu = 1.217 \times 10^{-5} \, \frac{ft^2}{sec}$$

Let V = average velocity of water in the pipe.

$$\mathbf{R} = \frac{\rho V D}{\mu}$$

$$= \frac{1.938 \times 4/12\, V}{2.359 \times 10^{-5}}$$

$$= 2.74 \times 10^4 V$$

for

$$h_T = 25 \text{ ft} = f\frac{L}{D}\frac{V^2}{2g} + (k_1 + k_2 + k_3 + k_4)\frac{V^2}{2g} + h$$

$$= f\frac{150}{4}\frac{V^2}{2g} + (10 + 2.06 + 10 + 5.5)\frac{V^2}{2g} + 0.2$$

$$24.8 = \frac{V^2}{2g}(37.5f + 27.56)$$

$$V = \frac{39.96}{\sqrt{37.5f + 27.56}}$$

then

$$\mathbf{R} = 2.74 \times 10^4 \frac{39.96}{\sqrt{37.5f + 27.56}}$$

$$= \frac{109.49 \times 10^4}{\sqrt{37.5f + 27.56}}$$

Assuming $f = 0.021$

$$\mathbf{R} = \frac{109.49 \times 10^4}{\sqrt{37.5 \times 0.021 + 27.56}} = 2.056 \times 10^5$$

with $\epsilon = e/D = 0.0026$

Entering the Moody diagram we get $f = 0.026$. Substituting this value into $\mathbf{R}$, we have

$$\mathbf{R} = \frac{109.49 \times 10^4}{\sqrt{37.5 \times 0.026 + 27.56}} = 2.049 \times 10^5$$

From the diagram again we obtain $f = 0.026$; therefore use $f = 0.026$. Hence,

$$V = \frac{39.96}{\sqrt{37.5 \times 0.026 + 27.56}}$$

$$= 7.48 \text{ fps}$$

Hence the discharge

$$Q = AV = \frac{\pi}{4}\left(\frac{4}{12}\right)^2 (7.48) = 0.65 \text{ cfs}$$

DESIGN OF PIPING SYSTEMS

To calculate the amount of energy input required in a pipeline, one has to take into account the friction loss, the local losses, and the energy required at the discharge point of the pipe. In most practical applications water is fed through pipes to several points from the same supply. This requires many pipe branches. In these problems the pipe friction losses are dominant.

The conveyance method is useful if the

problem involves several pipes as is true of branching pipes, as shown in Figure 4.5.

A problem of this type is analyzed by writing the continuity equation for the branching point, where

$$\sum Q = Q_1 + Q_2 + Q_3 = 0 \qquad (4.19)$$

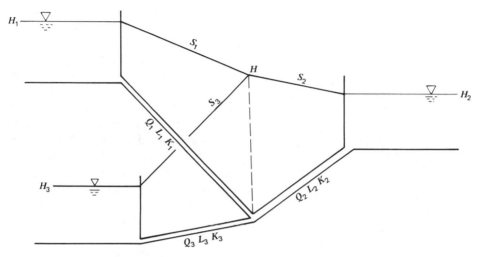

FIGURE 4.5 Branching pipes.

Since the pressure at the branching point must be single-valued, there exists a value of H, the energy at the branching point. The energy losses through each pipe are then

$$S_1 = \frac{H_1 - H}{L_1} = \left(\frac{Q_1}{K_1}\right)^2$$

$$S_2 = \frac{H_2 - H}{L_2} = \left(\frac{Q_2}{K_2}\right)^2 \qquad (4.20)$$

$$S_3 = \frac{H_3 - H}{L_3} = \left(\frac{Q_3}{K_3}\right)^2$$

Depending on the relative magnitude of H, the values of S may be positive or negative depending on whether the flow at the branching point is toward or away from the junction.

There could be a number of conditions for known or unknown parameters in such problems. The solution of any of these can be obtained by substituting the known terms into the preceding four formulas and solving them in a simultaneous man-

ner. By the basic concepts of algebra, for the four equations we may have four unknowns. These problems are usually solved by trial and error.

The principles used in the solution of branching pipes are also valid for the solution of pipe networks where there are scores of internal branching points. In municipal and industrial piping networks many interconnected pipes are involved. The number of the resulting nonlinear algebraic equations to be solved is large. To solve the multitude of interdependent equations in practical applications, electronic computers are utilized. There are several computer programs available for the design and analysis of complex networks.

To solve smaller network problems a systematic trial and error procedure developed by H. Cross may be used. The first step in the application of the Cross method is to assume initial values for either all discharges in the individual pipes or all piezometric heads at the junction points. The second procedure will be explained below.

Denoting the assumed initial piezometric heads at junction points A, B, C, etc., as H_A, H_B, H_C, ..., the corresponding discharges can be calculated by Equation

4.10 because the head loss h_2 for each pipe is given by the difference of heads at the junctions connected by the pipe in question. The conveyances of all pipes as well as their lengths are also known. After calculating the discharges for all pipes the continuity equation, Equation 4.19 must be written for all junctions. For the initially assumed piezometric heads to be correct, a remote possibility, all continuity equations must be satisfied. Otherwise, an error in the continuity equations, ΔQ, must appear at each junction in the form of

$$\sum Q = \Delta Q \qquad (4.21)$$

This indicates that the assumed piezometric heads must be adjusted. The amount of adjustment, ΔH, is done in a systematic manner.

By Equation 4.10 we may write, for a pipe of the network that

$$Q + \Delta Q = \left[\frac{K^2}{L}(h_L + \Delta H)\right]^{1/2} \qquad (4.22)$$

where ΔH is the adjustment in piezometric head at one end of the pipe. From Equation 4.22 it follows by rearranging and expanding the polynomial that

$$h_L + \Delta H = \frac{L}{K^2}[Q^2 + 2\Delta Q(Q) + \cdots]$$

Neglecting higher order terms and subtracting h_L from both sides, we obtain

$$\Delta H = \left(\frac{L}{K^2}Q\right)2\Delta Q \qquad (4.23)$$

Recognizing that the term in the parenthesis equals h_L/Q, we may put Equation 4.23 into the form of

$$\Delta H = \frac{2\Delta Q}{Q/h_L} \qquad (4.24)$$

Equation 4.24 establishes the relationship between the head adjustment at one junction point and the head loss and discharge in one connecting pipe. Obviously there are three or more pipes connected at any one junction point. The aggregate effect of the ΔH head adjustment on the excess discharge Q at a junction may be considered by the weighted formula

$$\Delta H = \frac{2\Delta Q}{\sum |Q/h_L|} \qquad (4.25)$$

where the denominator is the sum of all incoming and outgoing discharges divided by their respective head losses in the connecting pipes. After calculating all ΔQ values for the initially assumed piezometric heads, the required head corrections can be determined by Equation 4.25 at each junction point. For these new piezometric heads new head loss and discharge values may then be computed. The process is repeated until the errors in discharges are reduced to an acceptable level.

For the sake of convergence of the solution and to prevent excessive oscillations in the successive trial results, the factor 2 may be dropped from Equation 4.25. Equation 4.25 is not sacrosanct. The experienced user may adjust the H values in any arbitrary manner in order to obtain improved results. Note that only internal pressure heads are adjusted in the trial and error procedure. Exit and inlet discharge points may have their pressure values fixed throughout the computation depending on external controlling conditions. Example 4.9 indicates that the convergence of the solution is sometimes rather slow and the errors in the continuity equation written for all external (inflow and outflow) discharges may be considerable even after several trial solutions. This, and the sheer volume of the required computations make computer analysis desirable for network problems.

EXAMPLE 4.9

From a water tower at point A where the water elevation is 1000 ft above datum level a pipe network is fed as shown in Figure E4.9. The network is composed of steel pipes with an equivalent

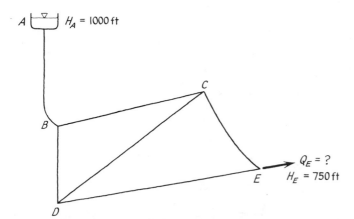

FIGURE E4.9

sand roughness of $e = 0.0005$ ft. The pipe lengths and diameters are tabulated in the solution. Assuming that at point E the exit pressure is maintained at 750 ft above datum level, calculate the discharge through the network using the Cross method.

SOLUTION:

First, compute the conveyance and the pipe resistance values for all pipes, shown in the first five columns of Table E4-9a. Next, assume initial pressure head values for all internal junction points as shown in the second column of Table E4-9b.

Table E4-9a

Line	L	D	K cfs	L/K^2	h_{L_I}	Q_I
AB	1000 ft	12 in.	60	0.277	100	19.0
BC	1500	8	17	5.19	50	3.10
BD	1200	12	60	0.333	100	17.33
CD	1500	6	7.5	26.66	50	1.37
CE	2000	12	60	0.555	100	13.42
DE	1000	8	17	3.46	50	3.80

Now in the first table show all head loss values for these trial pressure heads and then calculate the corresponding discharges, shown in the last column of Table E4-9a, by Equation 4.10.

Next calculate the excess discharges by Equation 4.21. Then compute the head corrections by Equation 4.25 using a reducing factor of 0.5 to avoid excessive oscillation in the trial solutions.

Detailed computations for the successive trials are not included.

Table E4-9b First Cycle of Computation

Internal Junction Points	Assumed H_1	$\Sigma Q =$	ΔQ	$\Sigma \lvert Q/h_L \rvert$	ΔH	$H_2 = H_1 + \Delta H$
A	1000	$+19 - 19$	0	—	—	1000^b
B	900	$+19 - 3.10 - 17.33$	-1.43	$\dfrac{19}{100} + \dfrac{3.10}{50} + \dfrac{17.33}{100} = 0.4253$	-3.36	896.64
C	850	$+3.10 - 1.37 - 13.47$	-11.69	$\dfrac{3.10}{50} + \dfrac{1.37}{50} + \dfrac{13.42}{100} = 0.2236$	-52.28	797.72
D	800	$+17.33 + 1.37 - 3.80$	$+14.9$	$\dfrac{17.33}{100} + \dfrac{1.37}{50} + \dfrac{3.80}{50} = 0.2767$	$+53.82$	853.82
E	750	$+3.80 + 13.42 - 19^a$	-1.78	—	—	750^b

[a] By continuity Q_E exit flow must equal Q_{AB}.
[b] Fixed elevations.

However the results of the first five cycles of computations are shown in Table E4-9c:

Table E4-9c

Junction Points	Initial H	H_2	H_3	H_4	H_5
A	1000	←		→	1000 ft
B	900	896.64	903.92	907.64	912.97
C	850	797.72	784.62	779.54	777.94
D	800	853.82	855.71	870.77	878.64
E	750	←		→	750

As shown above the solution is converging in a monotonous manner.

The discharge values for the fourth trial, however, still indicate a significant error as shown in Table E4-9d. By continuity Q_E exit flow must be equal to the flow in pipe AB, and also equal to the sum of the flows in CE and DE. The data indicates that additional trial solutions will be required before a correct solution is attained.

Table E4-9d

Pipe	Discharge cfs
AB	18.26
BC	4.96
BD	10.52
CD	1.84
CE	7.29
DE	5.90

WATER HAMMER

Depending on the relative position of the pipe with respect to the datum plane of the energy lines, the energy at any point of the line will be composed of the kinetic energy $(v^2/2g)$ and pressure energy (p/γ). Water will move through a pipe regardless of its position, as long as it is below the height representing the total available energy less the kinetic energy. The pressure on the pipe walls will be determined by the pressure energy term. Wall thicknesses of standard pipes are designed so that they can withstand ordinary service pressure. Sudden change of discharge may, however, result in stresses of sufficient magnitude to exceed these design stresses.

Any change of discharge in a pipe (valve closure, pipe fracture, pump stoppage, etc.) results in a change of momentum of the flow. By virtue of the impulse-momentum equation, this will cause an impulse force to be created, which is commonly called *water hammer*. This force should be checked as it may be detrimental to the system.

The theory of water hammer was developed by N. J. Zhukovsky*, as follows:

*Spelled Joukowski in older publications.

Consider a pipeline on which an open valve is located at a distance L (ft) downstream from a reservoir. Initially the fluid flows at a velocity v and the fluid pressure at the valve is p_0. If the valve is closed instantaneously, the fluid rams into the closed gate, decelerating to zero velocity and thereby creating a pressure shock. According to Newton, pressure shocks in fluids of infinite extent travel at a velocity given by the formula

$$c* = \sqrt{E/\rho} \qquad (4.26)$$

which is called celerity. E in this equation is the bulk modulus of elasticity (see Table 1-3) of the fluid (lb/ft^2); ρ is the density of the fluid (lb sec^2/ft^4). For some common liquids, c*, E, and ρ are given in Table 4-4.

If we compress the fluid in an elastic pipe, the pipe will expand. The modulus of elasticity E_c of a system composed of elastic fluid and elastic pipe may be calculated from the equation

$$\frac{1}{E_c} = \frac{1}{E} + \frac{D}{E_p w} \qquad (4.27)$$

where D is the pipe diameter, w is the thickness of the pipe wall, and E_p is the modulus of elasticity of the pipe material. Values of E_p for different pipe materials are listed in Table 4-5. Equation 4.27 refers to circular pipes only. The value of E_c for noncircular conduits is considerably smaller because of their limited structural rigidity under outward pressure.

The celerity of a shock wave in a pipe, c

Table 4-4 Celerity $c*$, Bulk Modulus of Elasticity E, and Density ρ for Some Common Liquids at 60°F

Liquid	$c*$ ft/sec	E lb/ft^2	ρ lb sec^2/ft^4
Water	4950	45×10^6	1.94
Seawater	4750	45×10^6	1.99
Benzene	3510	21×10^6	1.71
Crude oil	4600	35.9×10^6	1.70
Mercury	1460	56.2×10^6	26.3
Carbon tetra-chloride	3060	28.9×10^6	3.095

Table 4-5 Modulus of Elasticity, E_p of Various Pipe Materials.

Pipe Material	E_p (in million psi)	(in lb/ft^2)
Lead	0.045	6.48×10^6
Lucite (at 73°F)	0.4	57.6×10^6
Rubber (vulcanized)	2	288×10^6
Aluminum	10	1440×10^6
Glass (silica)	10	1440×10^6
Brass, bronze	13	1872×10^6
Copper	14	2016×10^6
Cast iron, grey	16	2304×10^6
Cast iron, malleable	23	3312×10^6
Steel	28	4032×10^6

fps, can then be computed from

$$\frac{c}{c^*} = \frac{1}{\sqrt{1 + ED/E_p w}} \qquad (4.28)$$

which is plotted in Figure 4.6. The graph indicates the considerable influence of the pipe rigidity on the velocity of the shock.

the valve is

$$t = \frac{L}{c} \qquad (4.29)$$

when the shock will disappear. At this instant the compressed, halted fluid in the pipe will not be balanced from that end.

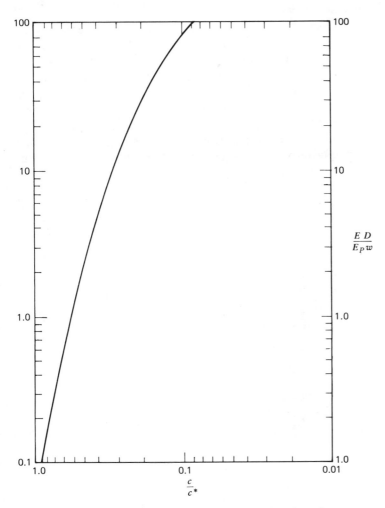

FIGURE 4.6 Celerity of pressure waves in pipes, c = celerity in elastic pipe. c^* = celerity in fluid of infinite extent.

The shock waves that travel upstream and downstream from the adjusted valve will ultimately reach the ends of the pipe where the pressures are controlled by stationary energy levels, for example, by reservoirs. The time t for a shock wave to reach such a point at a distance L from Hence, to relieve the compression, it starts to flow in the opposite direction. This creates a relief pressure shock that travels back to the valve. The time period T while the shock pressure acts on the valve is the time it takes for the pressure wave to travel from and back to the valve,

72 Flow in Pipes

that is,

$$T = 2t = \frac{2L}{c} \quad (4.30)$$

At this time all the fluid will be moving backward at a velocity v. Since the valve is closed, there is no supply for this flow; hence a suction (negative pressure shock) is created at the valve. This again will travel to and back from the reservoir, reversing the flow. Such oscillations of pressure and periodic reversal of flow will persist until its kinetic energy will be dissipated by friction. The process described will occur both upstream and downstream from the valve, the only difference being that the initial shock will be positive on the upstream side and negative

on the downstream side. The magnitude of the pressure shock at instantaneous valve closure is

$$p^* = \rho c v \quad (4.31)$$

and the pressure will oscillate in the pipe within the range of

$$p = p_0 \pm p^* \quad (4.32)$$

Either value could be detrimental to the pipeline. A nomographic solution of Equation 4.31 is shown in Figure 4.7.

The time of closure of a valve is really not zero but a certain finite period, say, T_c. The water hammer pressure increases gradually with the rate of closure of the valve. Depending on whether T_c is smaller or larger than T of Equation 4.30, one

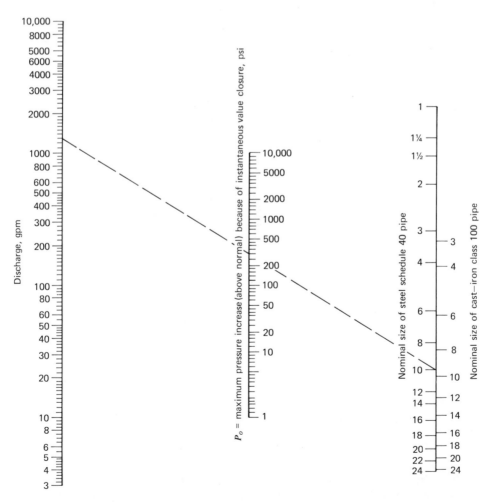

FIGURE 4.7　Water hammer nomograph.

differentiates between quick and slow closure.

If T_c is smaller than T, the shock pressure will reach its maximum value $p*$. Hence, quick closure is equivalent to instantaneous closure. If T_c exceeds T, $p*$ will not develop fully because reflected negative shock waves arriving at the valve after time T will counteract it. For such slow valve closures the maximum pressure may be calculated by the Allievi formula

$$p = p_0 \left[\frac{N}{2} + \sqrt{\frac{N^2}{4} + N} \right] \quad (4.33)$$

in which

$$N = \left(\frac{Lv\rho}{p_0 T_c} \right)^2 \quad (4.34)$$

Checking pipelines for water hammer effects involve the following procedures.

First, we determine the energy grade line and the hydraulic grade line for the piping system at the discharge flowing. On this basis we can locate the most critical points, where sudden change of discharge may occur (valves, pumps, etc.) and where bursting or collapse of the pipe may happen, for example, at low and high points. Also we may find the parameters $c*$ and E for the fluid from Table 4-4 and note the pipe size (D and w) and its elastic modulus E_p (Table 4-5). From this we may calculate the value of $ED/E_p w$. Using Figure 4.6 we obtain the system celerity, c. Now we may compute the wave travel times T from Equation 4.30 for both upstream and downstream directions. From operational considerations we assume the probable period of T_c time of closure. By the use of Equation 4.31 we may find the maximum and minimum pressures at any point where p_0 is known (from the hydraulic grade line). With particular attention to selected critical points we can check the pipe for bursting pressure or for cavitation and collapse by comparing p to the vapor pressure shown in Table 4-4. If $T < T_c$ the water hammer pressure will be reduced as computed by Equation 4.33.

Water hammer effects may be controlled by use of slow closing valves, standpipes, pneumatic shock absorbers, pressure relief valves, or similar devices.

EXAMPLE 4.10

When a 1000 ft long 12-in. diameter steel pipe with 0.375-in. wall thickness leads from a reservoir to a level 250 ft below the control gate is fully open, the water exits freely (Figure E4.10). Compute the

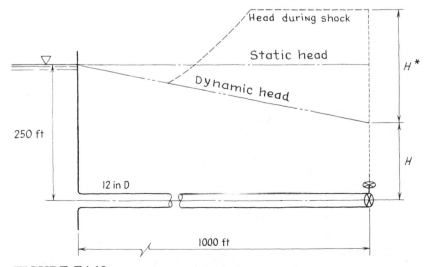

FIGURE E4.10

discharge and the head loss of the system. What is the maximum pressure if the valve is closed in $\frac{1}{2}$ sec, and how long will this pressure be sustained at the valve?

SOLUTION:

$$D = 12 \text{ in.} = 1 \text{ ft}$$

$$w = 0.375 \text{ in.}$$

$$e = 1.5 \times 10^{-4} \text{ ft}$$

and the kinematic viscosity is $\nu = 1.217 \times 10^{-5}$.

Let V = average velocity of water in pipe at the steady state.
Then, by writing Bernoulli's equation

$$\frac{P_A}{\gamma} + Z_A + \frac{V_A^2}{2g} + f\frac{L}{D}\frac{V_A^2}{2g} + \frac{k_1 V_A^2}{2g} = \frac{P_B}{\gamma} + Z_B + \frac{V_B^2}{2g}$$

in which $k_1 = 0.5$ \qquad (from Table 4.3)

$$0 + 0 + \frac{V^2}{2g}(1 + f1000 + 0.5) = 0 + 250 + 0$$

$$\frac{V^2}{2g}(1.5 + 1000f) = 250$$

$$V = \frac{126.88}{\sqrt{1.5 + 1000f}}$$

$$R = \frac{\rho VD}{\mu} = \frac{VD}{\nu}$$

$$= \frac{126.88 \times 1}{\sqrt{1.5 + 1000f}(1.217 \times 10^{-5})}$$

$$= \frac{104.256 \times 10^5}{\sqrt{1.5 + 1000f}}$$

Assume that $f = 0.013$.
Then

$$\mathbf{R} = 2.74 \times 10^6$$

From Equation 4.6

$$\epsilon = \frac{e}{D} = 0.00015$$

Entering the Moody diagram gave us $f = 0.013$. Therefore

$$V = \frac{126.88}{\sqrt{1.5 + 1000 \times 0.013}} = 33.32 \text{ fps}$$

The head losses of the system are friction loss and entrance loss:

$$h_L = f\frac{L}{D}\frac{V^2}{2g} + K_1\frac{V^2}{2g}$$

$$= \left(0.013 \times \frac{1000}{1} + 0.5\right)\frac{(33.32)^2}{2g}$$

$$h_L = 232.73 \text{ ft}$$

The steady state discharge $Q = AV$ is

$$Q = \frac{\pi}{4}(1)^2(33.32)$$

$$= 26.17 \text{ cfs}$$

If the valve is suddenly closed, the water hammer will affect the pressure in the pipe.

From Equation 4.26,

$$c^* = \sqrt{\frac{E}{\rho}}$$

where

$$E = 45 \times 10^6 \frac{\text{lbf}}{\text{ft}^2}$$

for water from Table 4-4

$$E = 1.94 \frac{\text{lbf sec}^2}{\text{ft}^4}$$

Then

$$c^* = \sqrt{\frac{45 \times 10^6}{1.94}} \text{ fps}$$

$$= 4816 \text{ fps}$$

From the celerity of shock wave in a pipe c fps, we get from Equation 4.28

$$\frac{c}{c^*} = \frac{1}{\sqrt{1 + ED/E_p w}}$$

(which may be solved by Figure 4.6 also).

where

E_p = Young's modulus of elasticity for pipe material

$$= 4.032 \times 10^9 \frac{\text{lbf}}{\text{ft}^2}$$

D = pipe diameter = 1 ft

w = pipe wall thickness = 0.375 in.

Then

$$c = \frac{4816}{\sqrt{1 + \frac{45 \times 10^6}{4.032 \times 10^9} \times \left(\frac{12}{0.375}\right)}}$$

$$= 4134 \text{ fps}$$

From Equation 4.29,

$$t = \frac{L}{c}$$

$$= \frac{1000}{4134} = 0.24 \text{ sec}$$

and $T_c = 2t = 0.48$ sec, which is larger than the valve closure time of

0.5 sec. Accordingly, we have quick closure. By using Figure 4.7 with $Q = 449(26.17) = 11,750$ gpm and a 12-in. steel pipe, we get

$P_{max} = 2000$ psi $= 1.7 \times 10^6$ lb/ft^2 above steady flow pressure level, sustained over 0.48 sec, which would certainly destroy the pipe unless the time of closure is significantly reduced.

Assuming $T_c = 3$ minutes, from Equation 4.33

$$P^* = P_0\left[\frac{N}{2} + \sqrt{\frac{N^2}{4} + N}\right]$$

in which

$$N = \left[\frac{LV\rho}{P_0 T_c}\right]^2 = \left[\frac{LV\rho}{\gamma H_0 T_c}\right]^2$$

$$= \left[\frac{LV\rho}{\rho g H_0 T_c}\right]^2 = \left[\frac{LV}{g H_0 T_c}\right]^2$$

where

$L = 1000$ ft

$V = 33.32$ fps

$H_0 = 250 - 232.73 = 17.27$ ft

$T_c = 180$ sec

so that

$$N = \left[\frac{1000 \times 33.32}{32.2 \times 17.27 \times 180}\right]^2$$

$$= 0.11$$

Hence

$$P^* = P_0\left[\frac{0.11}{2} + \sqrt{\frac{0.012}{4} + 0.11}\right]$$

$$\gamma H^* = \gamma H_0(0.391)$$

$$H^* = 17.27(0.391)$$

$$= 6.75 \text{ ft}$$

This is the pressure head rise at the valve. This head will be sustained at the gate for $2L/c = 2 \times 1000/4134 = 0.48$ sec.

OTHER UNSTEADY FLOWS

The physical mechanism explained in the previous section appears in connection with all cases of unsteady flows in pipes. To bring these ideas into focus, let us analyze what happens in a pipeline that connects a sump of a pumping station to a lake or reservoir. Consider the arrangement shown in Figure 4.8. Initially let us assume that the pump is not operating; therefore the water level in the sump is the same as in the lake. Hence the water is not flowing in the pipe. If we denote the pump capacity by Q_p, it will lift out that amount from the sump when started. When the pump is put into operation, the water will not yet flow in the pipeline. First, the water level in the sump will have to be reduced by pumping. If we know the surface area A_s of the sump with no inflow into it initially, the pump will reduce the sump's water level by a depth of h as

$$\frac{Q_p}{A_s} = \frac{h}{\Delta t} \tag{4.35}$$

during a period of Δt.

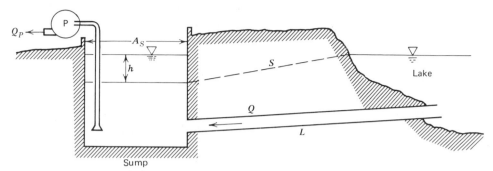

FIGURE 4.8   Feed line to sump.

Then, as soon as this reduction occurs the water in the pipe will have energy available to create flow corresponding to $H = h_L/L$ in which L is the pipe length. However, no fluid will flow into the sump until the water moves into the pipe from the lake. The "message" that energy became available at the sump will travel backward in the pipe as a pressure shock, just as it did for the water hammer. Accordingly, Equation 4.29 will come to bear. For long feed lines the travel time t of this shock wave may be of the order of magnitude of Δt. In this case by the time the water in the lake "learns" of the available energy h in the sump, the pump has already increased this depth considerably. After a time period of t the pipe will deliver a discharge Q. This discharge tends to fill the sump, and would cause a rise in the water level according to Equation 2.2a. If the pump ceased operating the pipe would fill up the sump gradually by a steadily decreasing discharge, because the available energy reduces as the water level in the sump increases. Because of the mechanism just described three things may happen. If the pump and the pipe length are large while the sump and the pipe diameter are small, the pump may empty out the sump and, by picking up air, the pump may stop operating. Conversely, if the pipe diameter and the storage capacity of the sump is large as compared to the pipe length and the discharge of the pump, the system will operate well although a large sump and a large pipe for a relatively small pump may be costly. In the intermediate condition the water level will fluctuate in the sump until a steady flow condition is reached. This is the case when the system is "properly tuned."

In the situation described above the unsteady flow in the pipe was controlled indirectly by the pump. When water flows out of a reservoir with a correspondingly decreasing amount of stored water the flow in the pipe is directly controlled by the changing elevation of the water in the reservoir. This problem is depicted in Figure 4.9. By continuity the discharge Q flowing in the pipe at any time t, when the water level in the tank is at Z_1 is

$$Q = V_r A_r = v_p A_p \qquad (4.36)$$

in which the subscripts r and p refer to the reservoir surface and the pipe, respectively.

By the energy equation

$$Z_1 = \left(1 + f\frac{L}{D}\right)\frac{V_p^2}{2g}$$

from which

$$v_p = n\sqrt{2gz_1} \qquad (4.37)$$

where n equals $1/\sqrt{1 + f(L/D)}$.

From Equations 4.36 and 4.37 the relationship between the water level in the reservoir and the time was derived by integral calculus, resulting in the formula

$$\Delta t = t_1 - t_0 = 2\,\frac{A_r}{A_p}\,\frac{1}{n\sqrt{2g}}\,(Z_0^{1/2} - Z_1^{1/2}) \qquad (4.38)$$

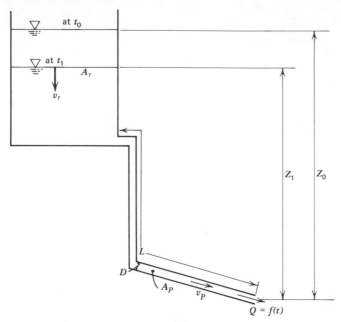

FIGURE 4.9   Drain line from storage tank.

The friction factor included in the term n can not be determined directly because when the velocity changes the Reynolds number also changes. However, it is satisfactory to compute the velocity in the pipe at an intermediate water level during the draining process by Equation 4.37 and determine the corresponding f value by using the method discussed in conjunction with Moody's diagram.

PROBLEMS

4.1   The allowable head loss in a very smooth plastic pipe is 40 ft. The pipe is 2000 ft long. Determine the diameter of the pipe if the discharge is 600 gallons per minute by using the Hazen-Williams formula.

4.2   The relative roughness of a pipe is 0.002. Determine the friction factor for a flow characterized by a Reynolds number of 2×10^4.

4.3   Check the result obtained for Problem 4.1 assuming that the water flowing is at 70°F. Use Li's diagram (Figure 4.3a).

4.4   For a galvanized iron pipe of 6 in. diameter calculate the slope of the hydraulic grade line if the discharge is 3 cfs. Use the conveyance formula (Equation 4.10).

4.5   A 1300 ft long, 8 in. diameter steel pipe is supplied at one end with 6 cfs of water. The pipe is perforated such that all the discharge is taken out uniformly throughout the length of the pipe. Assuming that the required piezometric pressure at the downstream end of the pipe is 15 ft of water determine the required pressure head at the inlet.

4.6   A 200 m long 4 in. steel pipe carries a discharge of 20 liters/s. The pipe includes a rotary valve closed 40 degrees, a fully open gate valve, and a perpendicular reentrant inlet. It discharges freely into the atmosphere. Compute the total head loss.

4.7   A 4-in. diameter steel pipe branching network connects three reservoirs. Each of the three pipes is 1500 ft in length. The water elevations in the three reservoirs are 1000, 900, and 850 ft above datum level. Determine the flow in each pipe.

4.8   A 1000 ft long steel pipe carries water from a reservoir with a water elevation at 682 ft to another pipe 265 ft above datum. The pipe has a diameter of 12 in. with a wall thickness of 2 in. A gate valve at the downstream end is slowly closed during a one-half minute period. Determine the maximum water hammer pressure.

4.9   A 4 m^2 sump is 3 m deep and serves a pump that moves water 50 liters/s when operating. The static water level in the sump is 2.5 m above the strainer bucket of the suction line of the pump and equal to the water level in a lake to which it is connected by a 6 in. steel pipe 200 m long. Analyze the conditions in the sump after the pump is turned on and determine whether the sump design is satisfactory.

4.10   A 15 ft diameter tank is full of water to a height of 8 ft. A 6 in. diameter steel drain pipe is carrying the water horizontally to a distance of 300 ft where it discharges freely. How long will it take to drain 80% of the stored water?

5

PUMPS

The proper selection of pumps is an integral part of the design of piping systems. Pumps are also often used in drainage, irrigation, and other areas, wherever it is required to add hydraulic energy to the water. In this chapter the basic hydraulic concepts of pumps will be outlined to provide sufficient understanding of their design, selection, operation, and maintenance.

PUMP CLASSIFICATIONS

Unless water is moved by gravity at an adequate discharge and pressure, it may be necessary to install pumps. Pumps add energy to water. There are many kinds of pumps used in the various technological fields. The three main classes into which pumps are sorted are centrifugal, rotary, and reciprocating pumps. These classes refer to different ways pumps move the liquid. The different classes could be further subdivided into pumps of different types. For example, centrifugal pumps include the following types:

Propeller (axial flow)
Mixed flow
Vertical turbine
Regenerative turbine
Diffuser
Volute

This classification of centrifugal pumps is based on the way the rotating component, the impeller, imparts energy to the water. In turbine pumps or radial flow pumps the impeller is shaped to force water outward at right angles to its axis. In mixed flow pumps the impeller forces water in a radial as well as in an axial direction. In propeller pumps the impeller forces water in axial direction only. Within radial flow pumps we speak of volute, diffuser, or circular type of casing, referring to the way water is collected and steered toward the exit pipe after it leaves the impeller.

Any one of the above types can be single stage or multistage pumps; stage refers to the number of impellers in a pump. Another distinguishing characteristic is the position of the shaft, which can be vertical or horizontal. There are single suction and double suction pumps. Pumps could also be grouped by their construction materials: bronze, stainless steel, iron, and their various possible combinations. Material becomes important when corrosive liquids are handled.

There are many other ways in which pumps may move liquids. However in the field of hydraulics, when we deal specifically with water, the most common pumps are centrifugal. Therefore our discussion will focus on these.

THE SPECIFIC SPEED

The selection of the type of pump for a particular service is based on the relative quantity of discharge and energy needed. To lift large quantities of water over a relatively small elevation—for example, in removing irrigation water from a canal and putting it onto a field—needs a different kind of pump than when a relatively small quantity of water is to be pumped to great heights, such as in furnishing water from a valley to a ski lodge at the top of the mountain. To make the proper selection for any application one needs to be familiar with the basic concepts of operation of the main types of centrifugal pumps.

The water enters the pump at the shaft that rotates the impeller. The impeller is a series of propellers, vanes, or blades that are arranged peripherally about the shaft and may or may not be held together by one or two circular plates. As the motor rotates the shaft at high speed exerting a torque T, the water is whirled around as it enters the impeller. The angular velocity ω imparted to the water particles throws them outward onto the wall of the casing. The casing is built so that it leads the water toward the exit pipe either by vanes or by its gradually expanding spiral shape. By proper design of the casing the kinetic energy imparted to the water by the impeller gradually changes into pressure energy.

Impellers of large radius and narrow flow-passages transfer more kinetic energy per unit volume than smaller radius impellers of large water passage. Pumps designed so that the water exits the impeller at a radial direction impart more centrifugal acceleration than those from which water exits axially or at an angle. Therefore, the relative shape of the impeller determines the general field of application of a centrifugal pump. Not that we need to take a pump apart to see the shape of its impeller before we select one for a particular use. Few people design pumps; those who use them generally deal with them on a "black box" basis. To simplify things the discharge, head, and speed at optimum performance of various pumps is consolidated into one number called *specific speed*. The specific speed n_s of a pump is computed from

$$n_s = \frac{\text{rpm}\sqrt{\text{gpm}}}{H^{3/4}} \qquad (5.1)$$

where n_s is the specific speed in revolutions per minute, rpm is the rotational speed of the shaft, gpm is the discharge in gallons per minute, and H is the total dynamic head in feet that the pump is expected to develop, all at optimum efficiency. Pumps built with identical proportions but different sizes have the same specific speed.

The specific speed of a pump is not really a speed in the physical sense, although it can be used in the sense that a pump reduced in size such that it would deliver one gpm to a height of one foot would run at its specific speed. In practice specific speed is only a number well suited to characterize the various types of centrifugal pumps. Generally, pumps with low specific speeds (500 to 2000 rpm) are made to deliver small discharges at high pressures. Pumps characterized by high specific speeds (5000 to 15,000 rpm) deliver large discharges at low pressures. The approximate relationships between specific speed, impeller shape, discharge, and efficiency are shown in Figure 5.1.

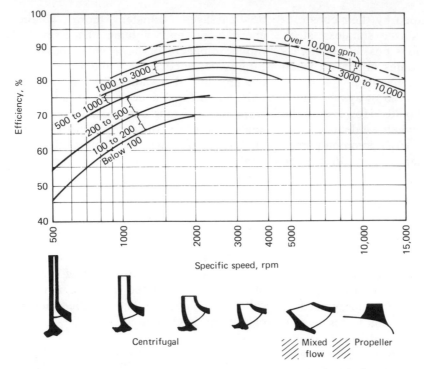

FIGURE 5.1 The relation of specific speed to the shape of the impeller, the discharge, and the efficiency of the pump. (Courtesy of Worthington Pump, Inc.)

EXAMPLE 5.1

Select the type of pump that is best suited for lifting 2000 gpm over a 6 ft tall dike when the available engine to drive the pump operates at 800 rpm.

SOLUTION:

For determining the optimum performance first, calculate the specific speed by substituting our data into Equation 5.1:

$$n_s = \frac{\text{rpm}\sqrt{\text{gpm}}}{H^{3/4}} = \frac{800\sqrt{2000}}{6^{3/4}} = 934 \text{ rpm}$$

From Figure 5.1 we find that we need a propeller-type pump that has a probable efficiency of 75%. From Figure 5.3 we find the approximate power need of the pump, which is 3 *HP*. Since this is the water horsepower we compute the brake horsepower by dividing it with the efficiency as follows:

$$\frac{3 \text{ water } HP}{75\% \text{ efficiency}} \times 100 = 4 \text{ brake } HP$$

Hence the available engine must develop a minimum of four horsepower.

HEAD AND POWER REQUIREMENTS

The dynamic head H to be developed by a pump is computed according to the methods described in Chapter 4. The pressure or head to be developed is the sum of the height to which the water is to be lifted from the level of the reservoir or sump from where the water is pumped. In addition to this height the friction losses occurring in the suction and discharge pipes must be included. Losses of energy through the inlet devices such as strainer and foot valve, elbows, valves and other components must also be taken into account. The pressure or kinetic energy required to be present at the end of the supply line is also a part of the *total* *dynamic head.* Pumps of the same type are manufactured in various sizes. The pump size is expressed by the diameter of its exit pipe. The approximate capacity in gpm can be determined by multiplying the square of this (inch) size by 25. Piping design should conform to this diameter, and the pipe size should be selected to minimize the initial installation cost as well as the capitalized cost of operation and maintenance. While a smaller diameter pipe is cheaper to install, the cost of the power required to overcome the larger friction losses throughout the life of the installation may be so high as to warrant its reduction by increasing the pipe size. The designer should strive to attain a judicious economical balance of these expenditures.

EXAMPLE 5.2

The 500 gpm is to be pumped to a height of 60 ft, overcoming all losses through a 156 ft long pipe that includes an open globe valve (Figure E5.2). Assuming that the motor is 3600 rpm, determine the pump size (Q, H, n_s) for the cases when the pipeline has a 6 or 8 in. diameter.

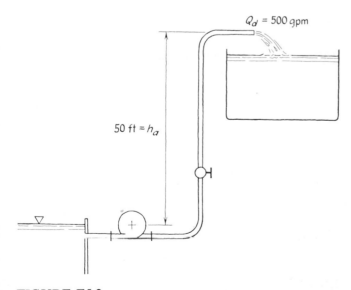

FIGURE E5.2

SOLUTION:

$$500 \text{ gpm} = \frac{500 \text{ gpm}}{450 \text{ gpm/cfs}} = 1.11 \text{ cfs}$$

$$\text{average velocity} = \frac{Q}{A} = V$$

For a 6 in. pipe,

$$\text{average velocity} = \frac{1.11}{(\pi/4)(0.5)^2} = 5.658 \text{ fps}$$

For an 8 in. pipe

$$\text{average velocity} = \frac{1.11}{(\pi/4)(8/12)^2} = 3.183 \text{ fps}$$

The total head loss equals friction loss in pipe plus friction loss in valve plus exit loss.

Consider a 6 in. pipe. Assume the water temperature is 65°F. From the properties of water at 65°F

$$\mu = 2.196 \times 10^{-5} \frac{\text{lbf sec}}{\text{ft}^2}$$

$$= 1.937 \frac{\text{slugs}}{\text{ft}^3}$$

$$\nu = \frac{\mu}{\rho}$$

$$= 1.134 \times 10^{-5} \frac{\text{ft}^2}{\text{sec}}$$

From Table 4-2 the roughness of the pipe (cast iron) = 0.00085 ft.

$$\frac{e}{D} = \frac{0.00085}{0.5} = 0.0017 \qquad \text{for a 6 in. pipe}$$

$$= \frac{0.00085}{8/12} = 0.00127 \qquad \text{for an 8 in. pipe}$$

Next calculate the Reynolds number as $\mathbf{R} = VD/\nu$.

$$\mathbf{R}_{6 \text{ in. } D} = \frac{5.658 \times 0.5}{1.134 \times 10^{-5}} = 2.49 \times 10^5 \qquad \text{for flow in a 6 in. pipe}$$

$$\mathbf{R}_{8 \text{ in. } D} = \frac{3.183 \times 8/12}{1.134 \times 10^{-5}} = 1.87 \times 10^5 \qquad \text{for flow in an 8 in. pipe}$$

By entering the Moody diagram, we get the following:

For a 6 in. D pipe condition $e/D = 0.0017$ and $\mathbf{R} = 2.49 \times 10^{-5}$; we obtain $f = 0.023$.

For an 8 in. D pipe condition $e/D = 0.0013$ and $\mathbf{R} = 1.87 \times 10^5$; we obtain $f = 0.022$.

Next calculate the friction loss in pipe h_L.

$$h_\text{L} = f \frac{L}{D} \frac{V^2}{2g}$$

For a 6 in. pipe 156 ft long,

$$h_{\text{L}_{6 \text{ in. } D}} = 0.023 \left(\frac{156}{0.5}\right)\left(\frac{5.658^2}{2g}\right) = 3.57 \text{ ft}$$

and for an 8 in. pipe 156 ft long,

$$h_{L_{8 in. D}} = 0.022 \left(\frac{156}{8/12}\right)\left(\frac{3.183^2}{2g}\right) = 0.81 \text{ ft}$$

The minor losses equal the loss in valve plus exit loss, which equal

$$k_1 \frac{V^2}{2g} + k_2 \frac{V^2}{2g}$$

or

$$(k_2 + k_2) \frac{V^2}{2g}$$

k_1 for globe valve = 10
k_2 for exit loss = 1

Then, minor losses in the 6 in. pipe $= (10 + 1) \left(\frac{5.658^2}{2g}\right) = 5.47 \text{ ft}$

$$8 \text{ in. pipe} = (10 + 1) \left(\frac{3.183^2}{2g}\right) = 1.73 \text{ ft}$$

Therefore the total loss in a 6 in. pipe equals friction loss plus minor loss.

$$H_6 = 3.57 + 5.47 = 9.04 \text{ ft}$$

Total loss in an 8 in. pipe is

$$H_8 = 0.81 + 1.73 = 2.54 \text{ ft}$$

Then the total discharge head

$$H_d = 60 + 9.04 = 69.04 \text{ ft for a 6 in. pipeline}$$
$$= 60 + 2.54 = 62.54 \text{ ft for an 8 in. pipeline}$$

Pump selection:
  From Equation 5.1 we determine the specific speed

$$n_2 = \frac{\text{rpm}\sqrt{\text{gpm}}}{H^{3/4}}$$

We get for a 6 in. pump:

$$n_s = \frac{3600\sqrt{500}}{(69.04)^{3/4}}$$
$$= 336 \text{ rpm}$$

and for an 8 in. pump

$$n_2 = \frac{3600\sqrt{500}}{(62.54)^{3/4}}$$
$$= 3619 \text{ rpm}$$

By entering into Figure 5.1 for $n_s = 3361$ and 3619 rpm, we note that the same kind of pump will suffice for both applications, a high specific speed radial flow pump with an approximate peak efficiency of 80%. Hence, we need a 6 in. pump having $Q = 500$ gpm, $H = 69$ ft, and $n_s = 3361$ rpm, or an 8 in. pump having $Q = 500$ gpm, $H = 62.5$ ft, and $n_s = 3619$ rpm.

The speed of the impeller is expressed in its rpm, meaning rotations per minute. Although some pump motors are built such that the speed can be varied, this is a somewhat expensive and rare case. For most pumps standard electric motors are used. Standard speeds of alternating current synchronous induction motors at 60 cycles and 220 to 440 volts are 3600, 1800, 1200, 900, 720, 600, and 514 rpm, depending on the number of poles. For a 50 cycle operation these speeds reduce to 3000, 1500, 1000, etc., rpm. Therefore the selection of the motor will determine the specific speed of the pump to a certain degree.

When selecting the motor for a pump and designing the wiring, fuses, and switching devices, it is important to know that pumps need more power for starting than during continuous operation. Since the full load speed of a regular electric motor is reduced by about 3 to 5%, the requirement of power increases considerably.

Power required by a pump may be computed from the formula

$$P = \gamma \cdot Q \cdot H \qquad (5.2)$$

where Q is the discharge in cfs, γ is the unit weight of the fluid in lb/ft³, H is the total dynamic head in ft. The dimension of P power is then given in ft-lb/sec. From mechanics we know that power may be expressed in terms of torque T and angular velocity ω

$$P = T \cdot \omega \qquad (5.3)$$

where T is (mass/time) · (radius) · (velocity). The angular velocity of the impeller can be computed from

$$\omega = \frac{2\pi}{60} \cdot (\text{rpm}) \qquad (5.4)$$

in radians per second.

Power is commonly expressed in terms of horsepower, HP, which, in English units requires that we divide the right side of Equation 5.2 by 550, that is

$$\text{water } HP = \frac{\gamma Q H}{550} \qquad (5.5)$$

which is called "water horsepower" because this is the power that water needs. For metric units Equation 2.7c may be used. Since pumps never perform at 100% efficiency, the water horsepower must be exceeded by a factor dependent on the efficiency of the operation. The power needed by the motor is called "brake horsepower," and may be computed as

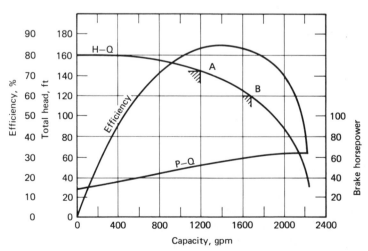

FIGURE 5.2   Typical pump performance graph. (From *Pump Application Engineering* by Hicks and Edwards, Copyright 1971, McGraw-Hill Book Co., Inc. Used with permission of McGraw-Hill Book Company.)

follows:

$$\text{brake } HP = \frac{\text{water } HP}{\text{efficiency}} \qquad (5.6)$$

Efficiency of a pump varies with Q and H. The value is included in the manufacturer's performance graphs available in pump catalogs, an example of which is shown in Figure 5.2.

The electric motor's power requirement must be expressed in kilowatts. One horsepower equals 0.745 kilowatt.

EXAMPLE 5.3

The pump characterized by known performance data (see Figure 5.2) is operated by a 50 horsepower electric motor. Determine the discharge and dynamic head delivered and the efficiency of the operation.

SOLUTION:

From the curve of P-Q obtained in Figure 5.2

$$Q = 1200 \text{ gpm}$$

Then from the efficiency curve at $Q = 1200$ gpm

$$e = 82\%$$

and from the head-discharge curve at $Q = 1200$ gpm,

$$H = 142 \text{ ft}$$

Pump selection in terms of the power required for a given discharge and dynamic head may be made by using the chart shown in Figure 5.3.

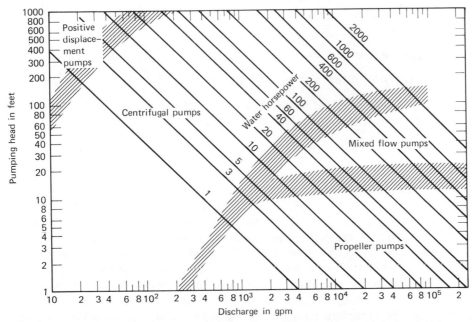

FIGURE 5.3   Power requirement for various discharges and heads of different types of pumps. (Courtesy of Fairbanks, Morse & Co.)

EXAMPLE 5.4

Determine the electric power requirement of a pump pumping 5000 gpm to a height of 30 ft if the overall efficiency is 80%.

SOLUTION:

The pump discharge 5000 gpm = $5000 \times 2.228 \times 10^{-3}$ cfs

$$= 11.14 \text{ cfs}$$

Thus the power used by the pump from Equation 5.2 is

$$p = \gamma \cdot Q \cdot H \left(\frac{\text{ft-lb}}{\text{sec}} \right)$$

where

$$\gamma = 62.4 \left(\frac{\text{lb}}{\text{ft}^3} \right)$$

$$Q = 11.14 \text{ cfs}$$

$$H = 30 \text{ ft}$$

Therefore

$$P = 62.4(11.14)(30) = 20{,}854 \left(\frac{\text{ft-lb}}{\text{sec}} \right)$$

Then water horsepower can be determined from Equation 5.5:

$$\text{water } HP = \frac{\gamma QH}{550} = \frac{20{,}854}{550} = 37.9$$

The overall efficiency of the pump is 80%; therefore the brake horsepower can be computed from Equation 5.6:

$$\text{brake } HP = \frac{\text{water } HP}{\text{efficiency}}$$

$$= \frac{37.9}{0.8} = 47.38$$

Since 1 horsepower = 0.7457 kW,

$$47.38 \ HP = (0.7457)(47.38) = 35.33 \text{ kW}$$

This is the electric power required to drive the motor of the pump and discharge 5000 gpm with a pressure corresponding to 30 ft of water.

PUMP SELECTION AND ALTERATION

In general hydraulic design work the selection of the proper type and size of pump and motor is left to the pump manufacturer. Catalogs of major manufacturers are available for the asking. Although these catalogs include complete data on the hydraulic performance criteria of pumps, such as the graph shown in Figure 5.2, most manufacturers supply a questionnaire on which the design requirements of a pump may be listed. From the data provided the manufacturer will advise the designer about the most suitable models. Many times some pumps are already available for the industrial designer in salvage yards or in other storage facilities. Similarly, the work may involve changes in existing plants because of increasing pumping needs. In these cases all attempts should be made to use these pumps, since pumps are expensive. Pump performance may be changed either by changing the impeller, changing the motor, or both. To change the pump performance characteristics certain basic laws valid for all centrifugal pumps are helpful. These are called *affinity laws* and are as follows:

Changing the impeller diameter D in the pump results in changes of Q, H, and P according to the relations

$$\frac{Q_1}{Q_2} = \frac{D_1}{D_2}$$

$$\frac{H_1}{H_2} = \left(\frac{D_1}{D_2}\right)^2 \qquad (5.7)$$

$$\frac{P_1}{P_2} = \left(\frac{D_1}{D_2}\right)^3$$

Subscripts 1 and 2 refer to values of the parameters before and after the change, respectively. When only the motor speed is changed on the pump, the resultant changes follow these relationships:

$$\frac{Q_1}{Q_2} = \frac{N_1}{N_2}$$

$$\frac{H_1}{H_2} = \left(\frac{N_1}{N_2}\right)^2 \qquad (5.8)$$

$$\frac{P_1}{P_2} = \left(\frac{N_1}{N_2}\right)^3$$

where N_1 and N_2 refer to the rpm of the motors before and after the change. The similarity in Equations 5.7 and 5.8 shows that the change of impeller has the same influence on pump performance as the change of speed.

EXAMPLE 5.5

A pump delivers 500 gpm at 70 ft dynamic head when run at 3600 rpm. Determine the change in operating characteristics if the pump's impeller is reduced from 6 to 4 in.

SOLUTION:

By Equations 5.7, with $Q_1 = 500$ gpm and $H_1 = 70$ ft,

$$\frac{500}{Q_2} = \frac{6}{4}$$

$$Q_2 = 333 \text{ gpm}$$

$$\frac{70}{H_2} = \left(\frac{6}{4}\right)^2 = 2.25$$

$$H_2 = 31 \text{ ft}$$

The required power will change to

$$P_2 = \frac{P_1}{(D_1/D_2)^3} = 0.29 P_1$$

From Equation 5.2

$$P_1 = \gamma Q_1 H_1 = 62.4(500)70 = 2.184 \times 10^6 \frac{\text{ft-lb}}{\text{sec}}$$

Therefore

$$P_2 = 0.29(2.184 \times 10^6) = 633,360 \frac{\text{ft-lb}}{\text{sec}}$$

THE ALLOWABLE SUCTION HEAD

An important point in the design of pumping installations is the elevation of the pump over the water level in the sump or reservoir from which the water is taken. The water in the suction line is in tension. The pressure is therefore lower than the atmospheric pressure. Adding to this already lowered pressure is the energy loss between reservoir and pump because of local losses and pipeline friction. The pressure is even more reduced because part of the energy at the pump is used in the form of kinetic energy, because of the high velocites in the pump casing particularly around the impellers. This latter effect is related to the specific speed of the pump. Adding the elevation of the pump, the kinetic energy head, and the friction losses in the suction pipe of a pump, one obtains the *total suction head* H_s. If this total suction head corresponds to a pressure reduction in the pump that is equal or less than the vapor pressure of the water (see Chapter 1) the water will change into vapor. This phenomenon is called *cavitation*. More than a half of the troubles experienced with centrifugal pumps can be traced to the suction side and many of these problems involve cavitation. If the water vaporizes in the pump small vapor bubbles form at the suction passages and at the impeller inlet. These bubbles collapse when they reach the region of high pressure. These collapses may occur at such violence that damage to the metal can result. Successive bubbles break up with considerable impact force,

causing local high stresses on the metal surfaces pitting them along the grain boundaries of the casing and at the tips of the impeller. The presence of cavitation is easily recognized; the vibration and noise make the pump sound as if it were full of gravel. The result of cavitation is a significant drop in efficiency and a subsequent mechanical failure of the pump because of the cavitational erosion of the casing and the impeller and fatigue failure of the seals and shaft.

To prevent cavitation the pump should be placed such that the suction head H_s is less than the head available, based on the atmospheric pressure minus the vapor pressure. The cavitation condition is described by the cavitation parameter

$$\sigma = \frac{\left(\dfrac{p_{atm}}{\gamma} - \dfrac{p_{vapor}}{\gamma}\right) - H_s}{H} \tag{5.9}$$

where the numerator is called net positive suction head, or NPSH. H in the denominator is the total dynamic head. Values of the σ cavitation parameter range from 0.05 for a specific speed of 1000 to 1.0 for a specific speed of 8000. The value of σ is usually furnished by the manufacturer on the basis of tests. For pumps of high specific speed, that is, low heads with large discharge capacity, the allowable net positive suction head may be less than zero, indicating that these pumps should be installed well below the reservoir water level in order to eliminate possible cavitation. In these instances the pump needs to be of the vertical shaft type so that the motor can be installed at an elevation above any possible flood levels.

EXAMPLE 5.6

Determine the allowable maximum suction head of a pump whose cavitation parameter is 0.25 if it operates at sea level and room temperature, when the total dynamic head is 140 ft.

SOLUTION:

From the definition of the cavitation parameter σ, Equation 5.9,

$$\sigma = \frac{\dfrac{p_{atm}}{\gamma} - \dfrac{p_{vapor}}{\gamma} - H_s}{H}$$

where

H_s = suction head in ft

γ = unit weight of water to be pumped

$$= 62.4\frac{lb}{ft^3} \qquad \text{at } 60°F$$

H = total dynamic head = 140 ft

$p_{atm} = 14.7 \times 144$

$$= 2116.8\frac{lb}{ft^2} \qquad \text{at sea level}$$

$p_{vapor} = 0.26 \times 144$

$$= 37.44\frac{lb}{ft^2} \qquad \text{at } 60°F$$

$\sigma = 0.25$

Therefore,

$$0.25 = \frac{\dfrac{2116.8 - 37.44}{62.4} - H_s}{140}$$

$$H_s = -1.67 \text{ ft}$$

Hence the pump inlet must be at least 1.67 ft below the water level in the sump.

Cavitation may also occur if the pump is not operating at or near its optimum point of its efficiency curve.

To begin the operation of a centrifugal pump it should be first primed, unless the pump is self-priming. Priming means to fill the pump with water, so that the impeller can create suction. Foot valves do serve the purpose of keeping the water in the pump between periodic operations, but often they leak. Priming requires that there be valves before and after the pump. These are to be closed before priming and starting the pump. After starting the inlet valve is always opened first. After a short period the outlet valve may be slowly opened. A closed outlet valve does not harm a centrifugal pump. With its outlet valve closed the pump pressure is in-creased by about 15 to 30%. The power load on the motor when the outlet is closed on the pump is reduced by about 50 to 60%.

TROUBLESHOOTING

Manufacture of pumps has reached such a level of sophistication that a pump is expected to give troublefree service for a long period of time. Troubles may arise from improper design or installation or from poor maintenance. Some common problems and their probable causes are listed below:

If no or not enough water is delivered:

Pump is not primed.

Speed is too low; check wiring.
Discharge head is too high.
Suction head is higher than allowed.
Impeller is plugged up.
Impeller rotates in the wrong direction.
Intake is clogged.
Air leak is in intake pipe.
Mechanical trouble (seals, impeller, etc.).
Foot valve is too small.
Suction pipe end is not submerged enough.

Not enough pressure:

Air is in the water through leak in suction pipe.
Impeller diameter is too small.
Speed is too low
The valve setting is incorrect.
Impeller is damaged.
Packing in casing is defective.

Erratic action:

Leaks are in suction line.
The shaft is misaligned.
Air is in water.

Pump uses too much power:

Too high speed.
Poorly selected pump.
Water too cold.
Mechanical defects.

Vibration and noise:

Cavitation.
Motor is out of balance.
Bearings are worn.
Propeller is out of balance (blade damaged).
Suction pipe picks up air through vortex action in sump.
Water hammer is in the piping system.

Noise due to cavitation can be eliminated by allowing some air to enter the suction pipe; however, this will not eliminate cavitation.

PUMPS IN PARALLEL OR SERIES

In pumping stations the discharge and head requirements may fluctuate considerably in time. To operate at optimum efficiency the output of a single pump may fluctuate only within a rather narrow limit. Hence, for fluctuating needs it is advantageous to install several pumps in a pumping station. Pumps installed together may operate in series or in parallel. In either case the individual performance characteristics of the pumps must be carefully matched in order that the best overall efficiency be attained. It is advantageous to install pumps of identical size. In this case their matching is best from hydraulic standpoints. Also, from a practical standpoint the cost of stocking spare parts will be less and interchangeability of the various components will facilitate repair and maintenance. To eliminate downtime for repair, an additional pump of similar size may be added also to the system. This is put on line when any one of the other pumps needs repair or maintenance.

When two identical pumps are installed together in series, as shown in Figure 5.4, the total output of the two is the same as the discharge of a single pump, but the output pressure is doubled. When connected in parallel, the total delivered discharge of two identical pumps is twice that of a single pump, but the output pressure remains the same as the single pump. The above concepts are true only if the pumps operating together discharge into the atmosphere. When delivering water into a piping system that offers frictional resistances, two pumps operating in parallel will encounter greater resistance to flow. Hence they will have a different operating point than when they operate alone in the same piping system. Figure 5.5 illustrates this concept. In this figure the discharge versus head curve is plotted for the piping system. The operating points for the single or parallel operation of two pumps are also shown. The

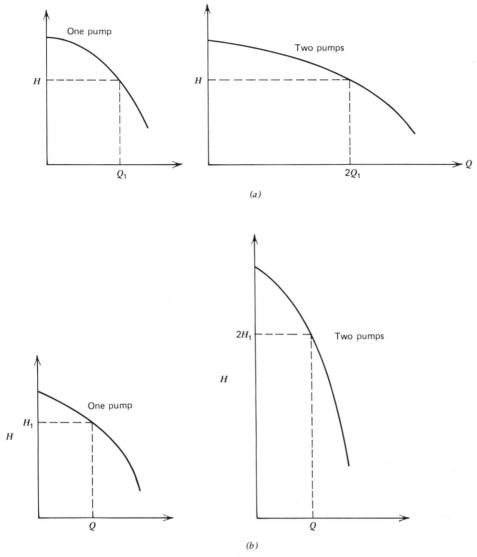

FIGURE 5.4   Two identical pumps operating in parallel (*a*) or in series (*b*).

intersection of these lines are the operating points. As indicated in the graph, the joint discharge of two pumps in parallel is less than twice the discharge of a single pump. Similarly, two pumps pumping in series will double neither head nor discharge in the whole system.

With these basic concepts, the joint operation of any arrangements of pumps can be analyzed. It requires the knowledge of the frictional resistances of the piping system and the availability of the characteristic curves of the pumps. The output of the whole system under various operating conditions can then be determined by graphical analysis.

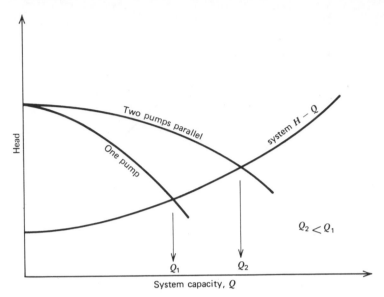

FIGURE 5.5   Operating characteristics of a piping system with one pump or with two pumps in parallel.

PROBLEMS

5.1   Calculate the specific speed of a pump that is operated by a 1800 rpm electric motor and delivers 50 gpm against a dynamic head of 45 ft. What type of pump will you select?

5.2   Calculate the torque required to operate the pump described in Problem 5.1. Neglect losses due to efficiency.

5.3   State the dynamic head, delivered discharge, efficiency and brake horsepower characteristics for both operating conditions, A and B, as shown in Figure 5.2.

5.4   For the pump with performance characteristics shown in Figure 5.2, which operate under condition A, determine the changes caused by changing the impeller from $D_1 = 6$ in. to $D_2 = 4.5$ in.

5.5   How would the torque requirement of the pump described in Problem 5.1 change by changing the driving motor to one running at 3600 rpm? What would happen to the discharge and head?

5.6   The cavitation parameter of a pump is 0.85 according to the manufacturer's data. Determine the allowable suction head for the conditions defined in Problem 1.2 if the total dynamic head is 56 ft.

5.7   The suction head of a pump having a specific speed of 3000 rpm is 6 ft. The total dynamic head is 40 ft, which includes 15 ft due to frictional losses in the piping system. It is found that at optimum operational conditions cavitation occurs. Recommend various possible remedial actions.

5.8   Select a pump that is best suited for the application described in Problem 4.5. State the specific speed, type, and other characterizing parameters.

5.9   Two pumps identical to the one for which the characteristic curves are given in Figure 5.2 operate at point *B* shown. If we assume that the pumps discharge into the atmosphere, what would be their combined discharge and head if they would be connected parallel or in series?

5.10   The two pumps described in Problem 5.9 are connected in series and deliver water into a horizontally laid 6-in. diameter plastic pipe that is 2000 ft long. At the end of the pipe the water discharges into a tank where the water level is 20 ft above the pipe. Determine the discharge.

5.11   Repeat Problem 5.9 for the case when the pumps operate in parallel.

SEEPAGE

The movement of water through the pores of soils often controls the safe and proper performance of earth works and hydraulic structures. Fundamental concepts of seepage are explained in this chapter. In addition, simple computational methods and numerous design aids are provided for the analysis of commonly encountered groundwater problems.

INTRODUCTION

The flow of water in the pores of the ground is usually an almost imperceptible process. For this reason the analysis of seepage is often relegated to secondary importance in the field of hydraulics. As the subject is inherently connected to soils, the concepts of groundwater flow are also considered to a certain extent in soil mechanics. As usual in such cases of common domain, seepage is rather inadequately covered in both hydraulics and soil mechanics books. Not that the problems arising from seepage are unimportant. In fact, the opposite is true. Often the effects of seepage determine whether a structure will stand or fail. Failures caused by seepage are often the costliest in terms of property and life. A collapsing trench during excavation or a failing dam or levee during floods all too frequently dominate the headlines of newspapers. Many of these problems as well as the less spectacular ones—landslides on slopes and flooding basements, for example—could have been avoided by a little attention to the concepts of seepage. This, of course, could come only with the fundamental understanding of these concepts.

PERMEABILITY AND SEEPAGE PRESSURE

Well over 100 years ago, Henry Darcy, a French water supply expert, explained the hydraulic behavior of sand filters used in the treatment of water. He postulated that the discharge Q of a sand filter of A surface and L thickness is proportional to the energy loss h that takes place across the filter. Figure 6.1 shows the arrangement considered by Darcy. The resulting

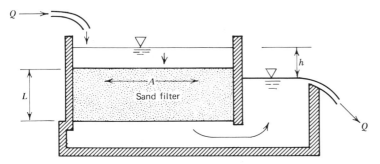

FIGURE 6.1 Darcy's sand filter experiment.

formula

$$\frac{Q}{A} = k\frac{h}{L} \qquad (6.1)$$

in which k, a constant of proportionality, was subsequently generalized for all cases of flow through porous materials in the form of

$$v = k\frac{\Delta h}{\Delta l} \qquad (6.2)$$

which is referred to as Darcy's law of seepage. The formula expresses that v at any point in the permeable material is proportional to the rate of loss in hydraulic energy at that point.

Darcy's constant of proportionality, k, is called the *coefficient of permeability* with the dimension of velocity. Later it was determined that k is really a compound of several physical parameters relating to both the fluid and the porous

solid. In fact

$$k = K'\frac{\gamma}{\mu} \qquad (6.3)$$

where K' now depends only on the sizes, shapes, and other geometric properties of the pores and minute channels in the soil, and γ and μ are the unit weight and viscosity of the fluid, respectively. Since the value of K' for soils is difficult to measure and we are almost always dealing with water of relatively uniform temperature, Equation 6.3 is used only to consider the effect of temperature variations on otherwise known conditions of seepage.

Typical values of Darcy's coefficient of permeability in different types of soils are shown in Table 6-1. The right column on this table shows that for different types of soils the order of magnitude of k varies by factors of ten. This means that the determination of the proper value of the per-

Table 6-1 Darcy's Coefficient of Permeability

Soil Type	Average Grain Size, mm	Range of k Coefficient, cm/s	Order of Magnitude of k, cm/s
Clean gravel	4–7	2.5–4.0	1
Fine gravel	2–4	1.0–3.5	1
Coarse, clean sand	0.5	0.01–1.0	10^{-1}
Mixed sand	0.1–0.3	0.005–0.01	10^{-2}
Fine sand	0.1	0.001–0.05	10^{-3}
Silty sand	0.02–0.1	0.0001–0.002	10^{-4}
Silt	0.002–0.02	0.00001–0.0005	10^{-5}
Clay	0.002	10^{-9}–10^{-6}	10^{-7}

EXAMPLE 6.1

A levee built of clay was placed on the top of a 6 in. sand layer. The bottom width of the levee is 60 ft. During flood the water elevation difference between the inside and the outside of the levee is 8 ft. If we assume that the permeability of the sand layer is 1.0 cm/s, calculate the expected seepage discharge for a 200 ft long portion of the levee.

SOLUTION:

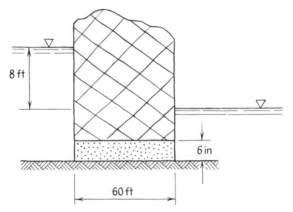

8 ft

6 in

60 ft

FIGURE E6.1

From Equation 6.1,

$$Q = Ak\frac{h}{L}$$

$$A = \frac{1}{2} \times 200 = 100 \text{ ft}^2$$

$$k = 1.0 \frac{\text{cm}}{\text{s}} = \frac{1}{2.54} \times \frac{1}{12} = 0.0328 \frac{\text{ft}}{\text{sec}}$$

$$h = 8 \text{ ft}$$

$$L = 60 \text{ ft}$$

thus

$$Q = 100 \times 0.0328 \times \frac{8}{60} = 0.437 \text{ cfs}$$

meability of a soil is at best an uncertain business. Laboratory evaluation of the k coefficients is standard practice, by using samples obtained from bore holes. This, in spite of its obvious scientific appearance, is a largely worthless operation. Seepage in most cases occurs in a horizontal direction; soil samples are tested usually for vertical flow. For geologic reasons most soil layers are far more permeable hori- zontally than vertically. Drilling and sampling of soils result in serious distur- bances to the soil. Sand and silt are greatly loosened in the process, and clay is com- pacted. Furthermore, since seepage oc- curs in huge masses of soil, samples are only minute parts of the whole mass. Claims that these samples are representa- tive of the whole are hardly convincing. By the nature of the formation of sands and gravels during the geologic processes one can expect considerable variations in permeability from point to point. Even pumping tests, another common way to arrive at a design value for permeability, give uncertain results. One thin gravel seam undetected during drilling can supply many times the discharge of the rest of the layers. The conclusion is that while all efforts must be made to obtain knowledge about the soils through which seepage will take place, the average value and variations of the permeability coefficient must be used with considerable suspicion until the project demonstrates its actual magnitude. For preliminary calculations values such as those in Table 6-1 may be sufficient, unless the magnitude of the project is such that the selection of permeability coefficients as well as the responsibility for such decisions can be placed on a reputable testing laboratory.

Although the topic of seepage immediately conjures up applications where the question of discharge is an important one, such as problems of dewatering, wells, and seepage losses from canals or reservoirs, there is a far greater number of cases in practice where it is of the least importance. Seepage is neglected almost always when the flow of water is not obvious; in these cases damage results at a later time. For example, consider clay saturated with flowing water. Because clay's very low permeability, flow of water to the surface is less than the rate of evaporation there, the clay surface may appear dry and the existence of seepage is disregarded. To look at the problem more closely, let us rewrite Equation 6.2 in the form

$$I = \frac{\Delta h}{\Delta l} = \frac{v}{k} \qquad (6.4)$$

in which the right side represents the ratio of two velocities. For clay the flow velocity is small because the permeability is also small. The ratio of two small numbers could be as much or more than that of two large numbers, even though for clays v may go unobserved.

The left side of Equation 6.4 represents the rate of loss of hydraulic energy along the paths of the seeping water. In the trench cut into saturated clay (Figure 6.2), the flow can be assumed to be nearly horizontal.

Taking a small cubic element of $(\Delta l)^3$ size in the clay mass shown on the left of Figure 6.2, we can recognize that the hydraulic energy lost from the seepage of the water through the element is Δh. This hydraulic energy is in fact transferred by viscous shear from the fluid to the soil. It manifests itself by an amount of pressure $\Delta p = \gamma \, \Delta h$ acting in the direction of the flow, distributed evenly in the element along the walls of its pores. The *seepage pressure* just described exerts a force on

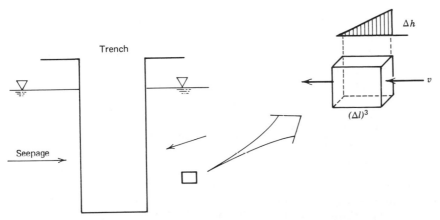

FIGURE 6.2   Seepage force in case of flow into a trench.

the soil at all points within the flow field. At the boundary of the flow field, along the sides of the trench shown in Figure 6.2, the seepage force acting on the soil tends to force the soil into the trench. This force is resisted by the internal strength of the soil, which for clay is its cohesion. Should the force of seepage be large enough to exceed the value of cohesion, the trench may fail from the effect of the seepage, although perhaps no water is observed in the trench. The seepage force, from the viewpoint of the soil mass, is a body force, similar to its weight. Its magnitude for an elementary cube of soil is obtained by multiplying the energy gradient $\Delta h/\Delta l$ acting on the cube with the unit weight of water and with the volume of the elementary cube. The direction of the seepage force is the same as the velocity of the water.

SEEPAGE ANALYSIS

In the last paragraph the hydraulic energy term Δh was dealt with in its elemental form. To solve actual problems encountered in practice such elemental notation is insufficient. One has to know the actual values to perform required computations. Herein lies the difficulty that makes the hydraulics of ground water flow a science in itself. To solve seepage problems one has to be able to define the physical boundaries within which seepage takes place. Once these boundaries are well-defined the values of the hydraulic energy must be known at the portions of the flow field where water enters as well as at portions where water exits.

By the law of conservation of energy, we know that

(energy at inflow) − (energy at outflow)

$$= \begin{pmatrix} \text{energy transferred to the soil} \\ \text{mass in the flow field.} \end{pmatrix}$$

The hydraulic energy is composed of only two parts: pressure energy (p/γ) and elevation energy (y). The third term, the kinetic energy ($v^2/2g$), is not considered since with the very small velocity values present in seepage the square of these velocities is a negligible amount. Therefore for seepage the hydraulic energy at any point is

$$h = \frac{p}{\gamma} + y \qquad (6.5)$$

where p is the pressure in the water, γ is the unit weight, and y is the elevation above a reference level.

For example, let us consider seepage from a lake into a nearby trench through a permeable sand layer that is bounded above and below by relatively impermeable clays (Figure 6.3). Depending on the water level kept in the trench, there are three distinct cases for the values of the hydraulic potential along the exit sections. Figure 6.3a shows the condition when the layer of sand is covered by water in the trench. Here the pressure at all points along B-C is defined by the depth of the point below the water surface, and the potential is a constant,

$$h_{\text{exit}} = (H - S) + y \qquad (6.6)$$

where the terms on the right are shown in Figure 6.3a. If the water is completely pumped out of the trench and the sand layer is exposed to air the pressure at all points on the surface is atmospheric, $p = 0$. In this case the hydraulic energy along B-C equals

$$h_{\text{exit}} = y \qquad (6.7)$$

a linearly varying term as shown in Figure 6.3b.

In the intermediate case when the layer is partially exposed the hydraulic energy at the exit is a combination of the two cases discussed above. This case is depicted in Figure 6.3c.

For all cases shown in Figure 6.3 the hydraulic energy at the inlet is constant with a value of H. This is because at points along A-D, as one goes downward the pressure grows at the same rate as the elevation decreases.

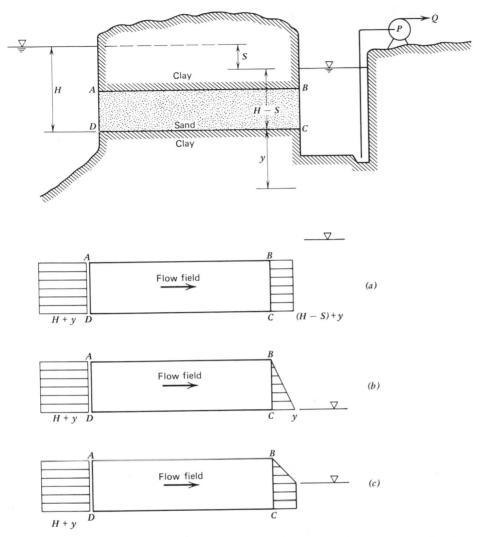

FIGURE 6.3   Entrance and exit potential of seepage through a confined permeable layer.

Let us now further consider the simple case of Figure 6.3a. The hydraulic energy causing the flow is the difference between the energy along A-D and the energy along B-C. This difference is denoted by S the drawdown in the trench, caused by the pumping. If the total length of the trench is disregarded by considering only a unit length as representative of the whole, the discharge per unit length, Q, by Darcy's equation (Equation 6.1), is

$$Q = Ak\frac{S}{L} \qquad (6.8)$$

where A is the cross-sectional area of unit length of the sand seam or simply the thickness of the seam if the problem is a two-dimensional one; L is the distance between lake and trench, that is, the length of the flow paths, and S is hydraulic energy available for flow. The formula above could be regrouped in the form of

$$\frac{Q}{kS} = \frac{A}{L} \qquad (6.9)$$

where the left side is called the relative discharge term and the right side represents the particular geometry of the flow field. Obviously the larger the cross-sectional area the greater the flow, and the longer the flow path the less the flow. The right side of Equation 6.9 can be generalized into a new term called "form factor," Φ. The concept of form factor provides distinct advantages for the solution of seepage problems, particularly for cases when two-dimensional representation is possible. Two-dimensional representation is valid in all cases when the flow field does not change regardless of where it may be considered. Seepage through levees, under long dams, into channels are the most typical of these problems. In two-dimensional cases Equation 6.9 takes a dimensionless form:

$$\frac{Q}{kS} = \frac{T}{L} = \Phi \qquad (6.10)$$

where Q is the discharge through the flow field of unit depth in the direction perpendicular to the flow field, and T is the "average" width while L is the "average" length of the flow field. Reasonably good approximate solutions for such seepage problems can be made very quickly by approximating the average measures of the width and the length of the flow field and dividing them to obtain the form factor.

EXAMPLE 6.2

A sand filter of the configuration shown in Figure E6.2 is 4 ft deep and its surface area is 40 ft². Determine the form factor of the flow field. If the permeability is 0.5 cm/s determine the discharge across the filter for a drawdown of 2.5 ft.

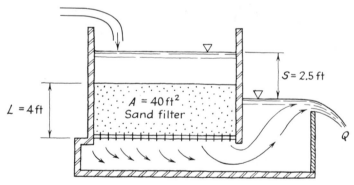

FIGURE E6.2

SOLUTION:

From Equation 6.9,

$$\frac{Q}{KS} = \frac{A}{L} = \Phi$$

where

$$A = 40 \text{ ft}^2$$

$$k = \frac{0.5}{2.54 \times 12} = 0.0164 \frac{\text{ft}}{\text{sec}}$$

$$S = 2.5 \text{ ft}$$

$$L = 4 \text{ ft}$$

then

$$\frac{A}{L} = \Phi = \frac{40}{4} = 10 \text{ ft}$$

Hence

$$Q = kS\Phi = 0.0164 \times 2.5 \times 10$$
$$Q = 0.41 \text{ cfs}$$

EXAMPLE 6.3

A concrete dam with 15 ft base width creates a 7 ft drop in the water surface. The dam is imbedded 2 ft into a 9 ft thick horizontal permeable layer that has a permeability coefficient of 0.1 cm/s. Estimate the seepage discharge under the dam by using the form factor method.

SOLUTION:

Draw the dam and the permeable layer to scale. Establish the approximate location of the average stream line length by roughly locating the position of the stream line representing 50% of the seepage flow above and below it. Find the distance A to B as shown in the sketch. To obtain the average width of the flow field we first recognize that it should be somewhat more than the depth of the layer below the dam since the part of the flow field before and after the dam is considerably wider, although portions far from the dam carry almost no discharge. The average width selected was between points C and D, the length of which was measured to be 8.5 ft. The distance between A and B is about 30 ft.

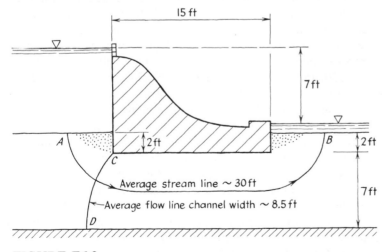

FIGURE E6.3

Accordingly, the two-dimensional form factor is

$$\Phi = \frac{8.5}{30} = 0.2833$$

By Equation 6.10 the relative discharge

$$\frac{Q}{kS} = \Phi = 0.2833$$

If

$$k = 0.1\frac{\text{cm}}{\text{s}} = 0.1 \times \frac{1}{2.54} \times \frac{1}{12}$$

$$k = 0.00328 \frac{\text{ft}}{\text{sec}}$$

$$S = 7\,\text{ft}$$

Hence

$$Q = \Phi \cdot k \cdot S = 0.2833 \times 0.00328 \times 7$$

$$Q = 0.00650\,\text{cfs per unit length of the dam}$$

Generally complete solution of a seepage problem involves the determination of the hydraulic potential values at all points of the flow field. The lines connecting all points having equal potential are called *equipotential lines*. The family of equipotential lines that are usually shown and labeled are the ones that represent each 10% or 25, 50, and 75% of the total energy, $H_{\text{in}} - H_{\text{out}}$, that causes the flow. Perpendicular to these equipotential lines are the *stream lines* of the flow. Stream lines are defined as lines that are tangential to the direction of the velocity of seepage at all points of the flow field. The family of stream lines that are usually shown and labeled are usually the ones dividing the total seepage discharge into percentages, for example, each 10%, or 25, 50, and 75% of Q. A plot showing equipotential and stream lines of a flow field is called a *flow net*. Flow nets are usually constructed in such a manner that pairs of adjacent equipotential lines and stream lines form a square. If the flow is curved these squares are distorted but retain their nature in the sense that at their corners the lines still cross perpendicularly and that a circle may be drawn into them in such a manner that it touches on each of their four sides. The form factor may be computed for such rectangular net by dividing the number of its equipotential lines by the number of its stream lines.

One of the most common methods for solving seepage problems is the drawing of flow nets. The graphical construction of flow nets usually start by drawing a few main stream lines (or potential lines) within the boundaries of the flow field. These, of course, are only rough approximations. Following this the other family of lines are drawn starting at one end of the flow field in such a manner that they form a proper "square" net. One must remember that the boundary lines are also parts of the net. Invariably the first trials result in incorrectly shaped "squares" or imperfect (non-perpendicular) intersections of lines. Hence the graphical process involves frequent erasures in parts of the flow net, until an approximately correct plot is produced. One will soon learn that even small local changes may cause significant variations in the whole plot. One of the most frequent mistakes made by beginners is the attempt to make the net too fine, drawing too many small squares. Starting with more than two or three stream lines in the beginning leads to this problem. A convenient process is to draw the fixed flow boundaries on an $8\frac{1}{2} \times 11$ in. paper, then to insert it in a transparent plastic cover (like Copco PS-5, for example). The trial flow net is then drawn on the plastic sheet where it can be erased and corrected with ease.

There are many ways to determine the form factor in a more precise manner for flow fields of different shapes, particularly

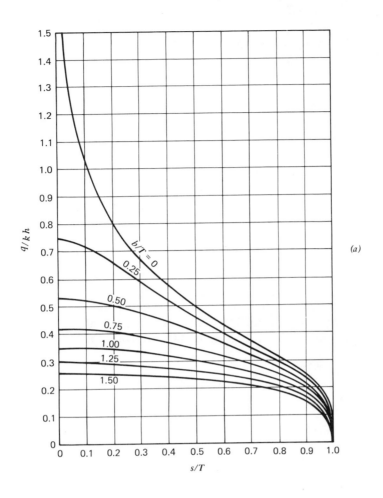

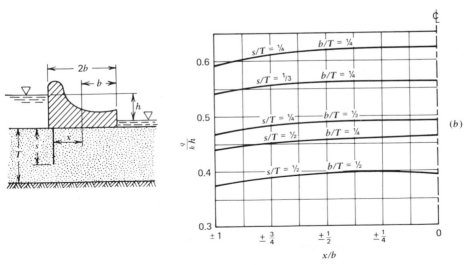

FIGURE 6.4 General solution for two-dimensional seepage under a hydraulic structure. (*a*) Sheet pile at center, $x = 0$. (*b*) Correction for offset sheet pile, $x \neq 0$. (From *Groundwater and Seepage* by M. E. Harr. Copyright 1962, McGraw-Hill Book Co. Inc. Used with permission of McGraw-Hill Book Company.)

under two-dimensional conditions. Most notable other methods include computer methods particularly those using relaxation techniques, analytic methods by using conformal transformation, and analog solutions involving electric and viscous flow analogies. Most of these methods are very cumbersome, complex, and theoretical, far beyond the scope of this book. Textbooks listed in the bibliography will provide ample guidance to those who are interested in these particular topics.

Fortunately for most practical cases solutions already exist in the extensive literature on seepage. For example, the case involving a dam with a sheet-pile underneath is reproduced in Figure 6.4. Figure 6.4 provides solutions for two extreme cases also. One is when the sheet pile stands alone, in which case b in Figure 6.4 equals zero. The other problem is when there is no sheet pile at all, that is, the depth of the sheet pile s is zero.

EXAMPLE 6.4

A 100 ft wide concrete dam with a sheet pile at the center is built over a horizontal silt layer 80 ft deep. The permeability of the silt is 10^{-5} cm/s. The head loss across the dam is 18 ft (Figure E6.4).

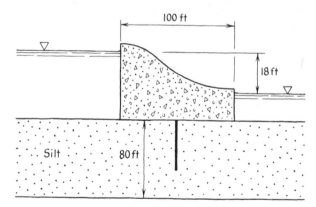

FIGURE E6.4

Determine the required sheet pile depth if the seepage discharge is to be reduced by 50% with respect to the case where no sheet pile exists. Also, determine if any further reduction of flow may be obtained if the sheet pile is placed at the front edge of the dam.

SOLUTION:

Consider Case I with no sheet pile:
From Figure 6.4,

$$S/T = 0$$

$$\frac{b}{T} = \frac{50}{80} = 0.625$$

Then

$$\frac{q}{kh} = 0.5$$

$$q = 0.5\left(\frac{10^{-5}}{2.54} \times \frac{1}{12}\right)(18)$$

$$= 2.95 \times 10^{-6} \text{ cfs/ft}$$

Case II: If the seepage discharge is to be reduced by 50% when the sheet pile is located at the center, then q/kh will be 0.25 and $b/T = 0.625$. From Figure 6.4 $S/T = 0.8$; therefore, the sheet pile depth $= 0.8 \times 80 = 64$ ft.

Case III: If the sheet pile is placed at the front edge of the dam, then $b/T = 0.625$, $S/T = 0.8$, and $q/kh = 0.25$, which is the same value as in Case II with the sheet pile at the center.

Note: From Figure 6.4, consider the value of q/kh. It is maximum when $x/b = 0$ or when the sheet pile is placed at the center. But in our case the value of $S/T = 0.8$ for which the relative discharge is so small that the influence of moving the sheet pile to the front does not seem to influence the solution.

EXAMPLE 6.5

Under the conditions described in Example 6.4 determine which method reduces the flow more: Increasing the width of the dam by 25 ft or increasing the sheet pile depth by 8 ft?

SOLUTION:

From Example 6.4 the sheet pile depth $= 64$ ft. If the sheet pile is increased by 8 ft, then $s = 64 + 8 = 72$ ft.

$$\frac{S}{T} = \frac{72}{80} = 0.9$$

If the width of the dam is the same, then $b/T = 0.625$. From Figure 6.4 the sheet pile at the center is

$$\frac{q}{kh} = 0.2 \tag{1}$$

and if the width of the dam increases by 25 ft,

$$2b = 125 \text{ ft}$$

So

$$\frac{b}{T} = \frac{125}{2 \times 80} = 0.78$$

the sheet pile depth $= 64$ ft, and $S/T = 0.8$. Thus from Figure 6.4, one again obtains

$$\frac{q}{kh} = 0.25 \tag{2}$$

Hence, we can see that by increasing the sheet pile depth we could reduce the flow more than by increasing the width of the dam.

The stability of a hydraulic structure encountering seepage underneath may be endangered by uplift pressures as the water exerts hydrostatic pressure at its base. The magnitude of this uplift pressure may be computed at any point of the base of the structure by Equation 6.5 in which y is the elevation of the point above the reference level and h is the hydraulic potential as determined by a flow net or some other means. The sum of all these hydraulic uplift pressures acting on the base of a structure is the uplift force. The magnitude and location of this force are important information in the determination of the required weight of a hydraulic structure. The stability of a structure is often endangered by excessive seepage

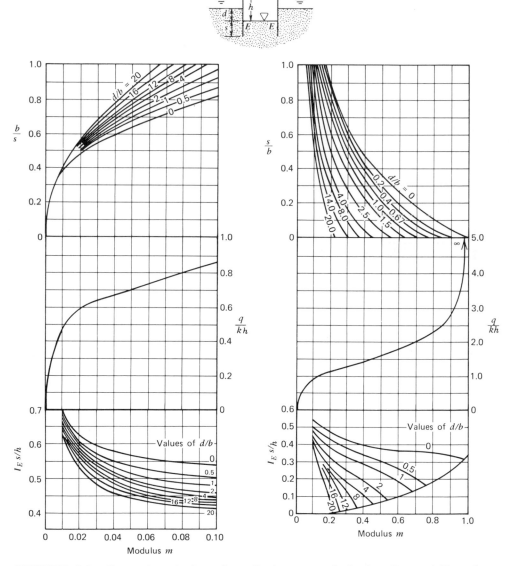

FIGURE 6.5 General solution for discharge and hydraulic stability for two-dimensional seepage into an excavation protected by sheet piles. (From *Groundwater and Seepage* by M. E. Harr. Copyright 1962, McGraw-Hill Book Co. Inc. Used with permission of McGraw-Hill Book Company.)

forces along the exit points of the flow. If the seepage force is large enough to pick up and carry away particles of the soil, erosion may occur. This phenomenon is often encountered with cohesionless granular soils subjected to seepage forces. In cases when actual erosion occurs (suffosion, that is, washing out of smaller soil grains of the soil matrix) we call it piping. If erosion of the grains does not occur but all strength of the soil is lost because of seepage forces present, it is referred to as "quick condition" (e.g., quicksand). In case of upward flow along a horizontal boundary of the field of seepage, the seepage force is resisted by the weight of the soil particles within the soil mass. Since the soil particles are submerged in water their actual weight is their net buoyant weight (see Figure 3.7) which equals the dry unit weight of the soil less the unit weight of the water. Because for most granular soils (gravel, sand, and silt) the dry unit weight is about twice that of water, their net buoyant weight is about the same as water's. On this basis we can define the hydraulic stability of a horizontal soil surface through which seepage takes place in an upward direction. Writing the difference of the net buoyant weight of the soil and the seepage force for a small elementary cube on the surface as

$$\frac{\Delta h}{\Delta l}\, \gamma_{\text{water}}(\Delta l)^3 - \gamma_{\text{submerged soil}}(\Delta l)^3$$
$$= \text{net stabilizing weight}$$

and by rearranging and substituting Equation 6.4 and recognizing that dry unit weight of most soils is about twice the unit weight of water, we obtain

$$\frac{v}{k} - 1 = \text{net stabilizing weight}/\gamma_{\text{water}}(\Delta l)^3$$

$$(6.11)$$

The critical condition is reached when the soil on the surface loses its net stabilizing weight, that is, the soil particles subjected to the upward seepage force appear to be weightless in the water. This requires that the energy gradient $\Delta h/\Delta l$ be unity at the surface; in other words, the discharge velocity v should be equal to the permeability of the soil or, in equation form,

$$v_{\text{critical}} = k$$
$$(6.12)$$
$$I_{\text{critical}} = 1$$

where I_{critical} is the gradient $\Delta h/\Delta l$ at the exit surface. A pertinent problem often found in practice is an excavation surrounded by sheet piles. Results of a complex mathematical solution of this problem are plotted in Figure 6.5. This plot gives the exit gradient for the flow under various conditions of excavation depths, sheet pile depths, and other pertinent geometric parameters.

EXAMPLE 6.6

Determine the critical gradient for the excavation in a river under the protection of two parallel sets of sheet piles 200 ft apart, if the excavation is performed in 12 ft of water to a depth of 10 ft below water and if the sheet piles are driven to a depth of 30 ft below the depth of the river (Figure E6.6). The soil has a permeability of 10^{-4} cm/s. Determine the required pump capacity for each 100 ft length of the excavation and determine the factor of safety with respect to piping or quick condition.

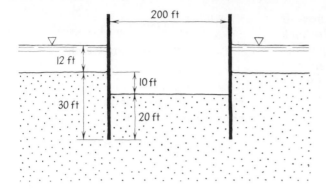

FIGURE E6.6

SOLUTION:

The variables are

$$h = 12 \text{ ft} + 10 \text{ ft} = 22 \text{ ft}$$
$$s = 20 \text{ ft}$$
$$d = 10 \text{ ft}$$
$$b = 100 \text{ ft}$$

Hence

$$\frac{s}{b} = 0.20 \qquad \frac{d}{b} = 0.1.$$

From Figure 6.5 $q/kh = 1.8$, where q is for 1 ft of length:

$$k = 10^{-4}\frac{\text{cm}}{\text{s}} = 10^{-4} \times \frac{1}{2.54} \times \frac{1}{12} = 3.28 \times 10^{-7}\frac{\text{ft}}{\text{sec}}$$

For 100 ft of length:

$$Q = 100q = 180kh = 180 \times 3.28 \times 10^{-7} \times 22$$
$$Q = 1.3 \times 10^{-3} \text{ cfs} = 0.581 \text{ gpm}$$

Again, from Figure 6.5 we obtain

$$I_{\text{exit}}\frac{s}{h} = 0.328$$

Hence

$$I_{\text{exit}} = 0.328 \times \frac{h}{s}$$
$$= 0.328 \times \frac{22}{20}$$
$$I_{\text{exit}} = 0.3608$$

Since by Equation 6.12 $I_{\text{critical}} = 1$. The factor of safety is

$$\frac{I_{\text{critical}}}{I_{\text{exit}}} = \frac{1}{0.3608} = 2.77$$

In many construction operations the flow of ground water results in the need of pumping. While this is sometimes considered a nuisance, the flow of ground water represents a certain factor of safety. Figure 6.6 shows two cases of identical geometry. In Figure 6.6a the impervious shale layer is left untouched. As there is no way for the water to flow upward into the excavation there is no loss of pressure. As a result the water pressure is allowed to build up to the level of the outside water elevation. This condition has caused catastrophic failures in excavation.

In Figure 6.6b there are weep holes drilled into the shale allowing the water to flow out of the layer. While this results in the need for continuous pumping, the

water pressure under the excavation is reduced through seepage losses, a factor that prevents excessive pressure to build up under the excavation.

FREE SURFACE PROBLEMS

For all cases considered so far the flow field was predetermined because the uppermost flow line was always bounded by an impervious boundary; therefore all parts of the flow moved at pressures larger than atmospheric pressure. These problems are called confined flow problems. But many practical problems are such that there exists a free surface bounding the flow field from the top. These are the unconfined flow cases.

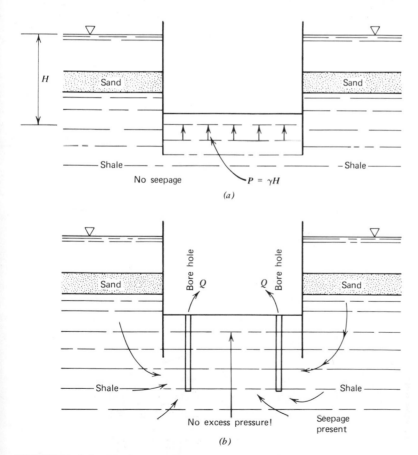

FIGURE 6.6 Hydraulic stability with an undrained (a) and drained (b) impervious layer at the bottom of an excavation.

112 Seepage

Some examples for unconfined flow are shown in Figure 6.7. Solving unconfined flow problems are even more difficult than those involving confined flow. The problem is that the location of the free surface is unknown; therefore the shape of the flow field, the form factor, is uncertain.

Graphical construction of flow nets for two dimensional free surface problems is possible but cumbersome. It starts with the drawing of a trial free surface which then defines the boundaries of a trial flow field. Construction of a flow net is then performed as explained before. Checking whether the trial free surface was drawn correctly is based on the fact that it is the limiting stream line on which the pressure is zero, hence values of the potential lines are, according to Equation 6.5, always equal the elevation of the free surface. Another requirement is that the potential lines must be perpendicular to the free surface. If these conditions are not satisfied by the trial free surface a new one must be drawn for which the process is repeated.

Solving problems of unconfined flow in an exact manner is complicated and the methods will not be explained in this book. Solutions of a few very commonly occurring problems with regard to flow through levees, earth dams, porous walls,

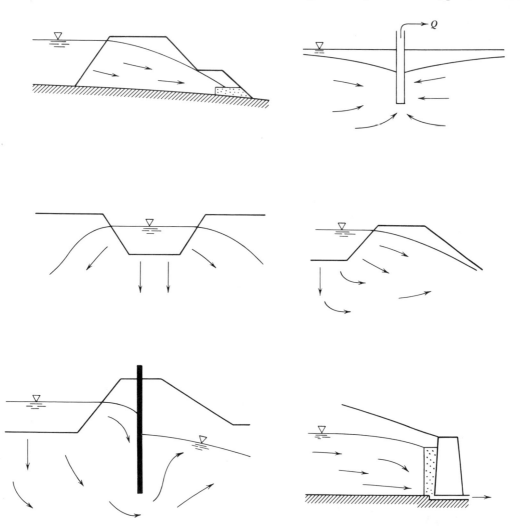

FIGURE 6.7   Examples of unconfined seepage.

and the like may be obtained by using the graphs shown in Figure 6.7. The graphs show the solution for flow through a porous vertical wall. This solution applies well in the case of a rock fill dam or earth dam resting on an impermeable base and built with a core of low permeability. The core in this case can be considered as a vertical wall and the remainder of the structure is disregarded because of its relatively high permeability.

EXAMPLE 6.7

A 30 m wide porous wall has a permeability of 10^{-3} cm/s. On one side the water level is 20 m; on the other side it is 1 m (Figure E6.7).

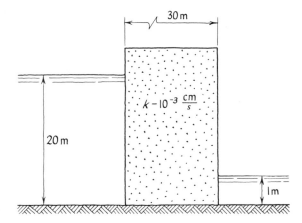

FIGURE E6.7

Determine the discharge for a 50 m long section of the wall and locate the free surface elevation at the downstream side of the wall.

SOLUTION

Porous wall

$$L = 30 \text{ m}$$

$$k = 10^{-3} \frac{\text{cm}}{\text{s}}$$

$$h = 20 \text{ m}$$

$$h_0 = 1 \text{ m}$$

We have

$$\frac{L}{h} = \frac{30}{20} = 1.5$$

$$\frac{h_0}{h} = \frac{1}{20} = 0.05$$

From Figure 6.7 with k/h and h_0/h above obtained, $S/h = 0.2$, $q/kh = 0.32$; therefore, the free surface elevation $S' = 0.2 \times 20 = 4$ m over the downstream level, and $q = 0.32 \times 10^{-3} \times 10^{-2} \times 20 =$

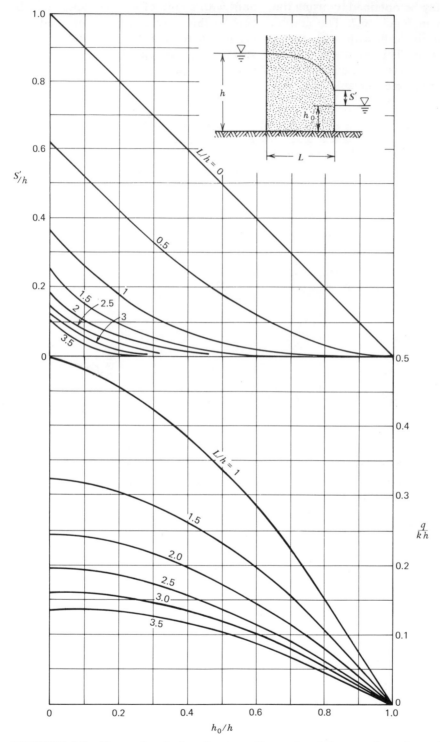

FIGURE 6.8 General solution for two-dimensional free surface flow through porous walls. (From *Groundwater and Seepage* by M. E. Harr. Copyright 1962, McGraw-Hill Book Co. Inc. Used with permission of McGraw-Hill Book Company.)

6.4×10^{-5} m^3/s for 1 m of wall. For a 50 m long wall

$$Q = 6.4 \times 10^{-5} \times 50$$

$$Q = 3.2 \times 10^{-3} \frac{m^3}{s}.$$

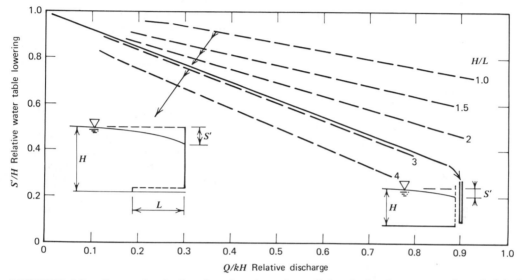

FIGURE 6.9 General solution for the elevation of the drained water surface behind retaining walls.

Figure 6.8 indicates that the exit point of the free surface at the downstream side is largely independent of the location of the downstream water level. This is logical because for all seepage flows the discharge will be the optimum under given energy conditions. If the free surface is lower than necessary, the form factor becomes smaller because of the smaller average thickness of the flow field. Since nature tends to deliver the most for the least, the flow field will take the shape of whatever would provide the optimal discharge for the given energy. Therefore the position of the exit point of the free surface S' will be somewhere above the downstream water level.

This principle is important in practice when the hydraulic performance of drains behind retaining walls is considered. Figure 6.9 shows the results of involved mathematical computations leading to a general solution of horizontal and vertical retaining wall drains. In all cases it is

assumed that the drains allow no downstream water level outside of the soil, that is, there is perfect drain operation. Hence the drawdown is complete, $S = H$. Figure 6.8 shows that, regardless of the design of the drains, the water table cannot be lowered by much more than 50% of the still water table. Only drain lengths of about four times the height of the water table extending horizontally into the soil provide less than this amount of lowering, a rather impractical design.

DISCHARGE OF WELLS

Formulas to determine the discharge of wells were first developed by J. Dupuit more than a century ago. Wells extract water from permeable water bearing layers called *aquifers*. One may distinguish between shallow wells that usually extract water from layers characterized by a free surface, and deep wells taking water from a buried aquifer in which water is under hydrostatic pressure. The latter are called *artesian* aquifers.

Shallow wells may be vertical that are dug, driven, or drilled, or horizontal, such as *galleries* or *collector pipes* radially arranged around a vertical concrete shaft. The latter are called horizontal collector wells (Ranney wells) a typical design of which is shown in Figure 6.10.

With certain theoretical simplifications Dupuit derived the discharge formula of a *shallow vertical well* under steady flow conditions, extracting water from a horizontal water bearing layer of infinite extent. His formula is commonly written as

$$Q = \frac{k\pi(H^2 - h^2)}{\ln(R/r)} \qquad (6.13)$$

in which the variables are as shown in Figure 6.11. The term R in Equation 6.13 is called the "radius of influence," an empirically determined factor that represents the distance from the well at which the free surface height remains unchanged (H) while the well is being pumped. One of the empirical equations known in the literature for the radius of influence is Sichardt's, which expresses R as a function of the drawdown in the well, S, and the permeability of the soil, in the form of

$$R = 3000S\sqrt{k} \qquad (6.14)$$

where k is to be substituted in meters per second. This formula is dimensionally incorrect and may be used for rough estimates only.

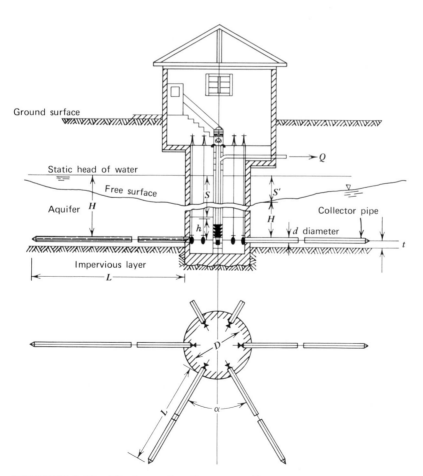

FIGURE 6.10   Horizontal collector well.

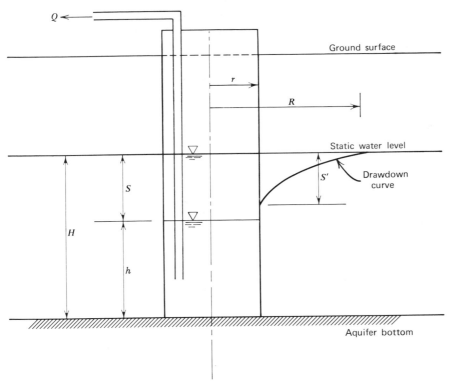

FIGURE 6.11 Interpretation of the variables for a well in unconfined horizontal aquifer.

The drawdown S is commonly measured inside of the well and is generally larger than the drawdown of the free surface level just outside of the well casing or screen. This situation is analogous to the one shown in Figure 6.9 except that here there is radial symmetry in the flow field instead of the two-dimensional case shown in the graph referred to. Laboratory experiments using sand models have shown that the lag between S and S'—the drawdown in the soil—depends on the relative drawdown, S/H, and also depends on the relative size of the well radius. Figure 6.12 shows the results of these laboratory studies. These results are most important because to use Dupuit's equation (Equation 6.13) correctly one has to substitute the drawdown in the water-bearing layer rather than where it is commonly measured, inside of the well. It is indeed fortunate that for small drawdowns in larger well diameters the lag

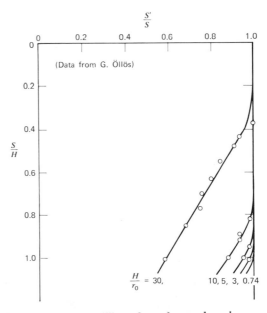

FIGURE 6.12 The drawdown lag in a shallow well. Experimentally measured drawdown characteristics of vertical wells with free surface.

between S and S' is not significant. But with small drilled wells, on the other hand, the error introduced could be quite considerable. The drawdown lag shown in Figure 6.12 (as well as in Figures 6.8 and 6.9) again shows Nature's optimization principle in action. The available energy, hence the velocity and discharge, grows with the drawdown. But as the drawdown increases the available flow area decreases. This in turn increases the velocity, which brings about an increase in energy losses. For a given discharge, S' represents the most economical energy utilization in the flow field.

EXAMPLE 6.8

A 24 in. diameter well is drilled into a shallow unconfined sandstone aquifer to its bottom 72 ft below the surface. From past experience the permeability of the sandstone may be estimated at 10^{-4} cm/s. Compute the discharge of the well when the static water level is 45 ft below the surface and the drawdown during pumping reduces the water level in the casing to 65 ft.

SOLUTION:

The variables are

$$H = 72 - 45 = 27 \text{ ft}$$
$$h = 72 - 65 = 7 \text{ ft}$$
$$S = 27 - 7 = 20 \text{ ft}$$
$$k = 10^{-6} \frac{\text{m}}{\text{s}} = 3.28 \times 10^{-6} \frac{\text{ft}}{\text{sec}}$$

By Equation 6.14

$$R = 3000 S \sqrt{k}$$
$$= 3000(20)\sqrt{10^{-6}} = 60 \text{ ft}$$
$$r = 1 \text{ ft}$$

Using Figure 6.12 with

$$\frac{S}{H} = \frac{20}{27} = 0.74$$

and

$$\frac{H}{r} = \frac{27}{1} = 27, \text{ say, } 30$$

we obtain $S'/S = 0.8$ from where $S' = 0.8 \times 20 = 16$ ft. For correct results $H - S' = 27 - 16 = 11 = h_*$ should be used in Dupuit's equation. Substituting into Equation 6.13, we get

$$Q = \frac{k\pi(H^2 - h_*^2)}{\ln(R/r)} = \frac{3.28 \times 10^{-6} \times \pi(27^2 - 11^2)}{\ln(60/1)}$$

$$= 0.00153 \text{ cfs} = 0.689 \text{ gpm}$$

Hence the capacity of the required pump shall be 0.689 gpm with a head of $72 - 20 = 52$ ft plus the required pressure at the well head.

The discharge formula for shallow wells, given as Equation 6.13 refers to steady flow from a well drilled into an aquifer characterized by a free surface. It is important to remember that the permanent discharge of a well can be no more than what is allowed by the discharge capacity of the aquifer. The permanent yield Q_p that may be obtained from a well is usually obtained by *pumping tests*. There are standard methods known in the literature for pumping tests of wells. Most of these are time-consuming and tedious procedures. From a practical standpoint two methods may be mentioned. One is to pump the well at a gradually stepped up discharge over an extended period of time, during which time the changes in drawdown are recorded. The maximum permanent yield of the well is the largest discharge that can be pumped indefinitely without any appreciable increase in drawdown and without a permanent inflow of fine sand and silt particles from the soil matrix. Another purpose of such a pumping test is to remove fine soil particles from the vicinity of the well that are not required to maintain the structural stabil-

ity of the soil but that actually reduce the permeability if left in place. For this reason the discharge levels during pumping are maintained until the water flows clean of silt.

Another method that may be used to determine the permanent yield in an indirect but quick manner is the Ponchet method. This calls for pumping the well at a constant discharge q for a short period of time, during which the gradual increase of the drawdown is carefully measured. This drawdown measurement is then continued after the pumping ceased, by which the rate of refill is established. The data is then plotted on a graph (Figure 6.13) in the form of time versus drawdown. The permanent yield Q_p is obtained by drawing tangents to the drawdown and refill lines at the peak point of the graph, and expressing the ratio of the slopes of the two tangents from where

$$Q_p = q \frac{\overline{ab}}{\overline{ac}} \qquad (6.15)$$

in which q is the constant pumping discharge applied and $\overline{ab}$ and $\overline{ac}$ are as shown in Figure 6.13.

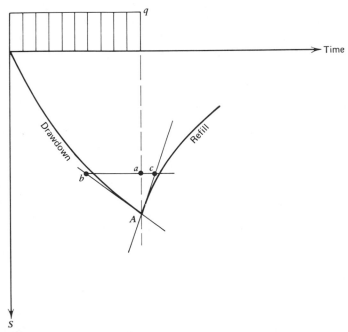

FIGURE 6.13   The Porchet method of analysis for well-pumping tests. $Q_p = q(\overline{ab}/\overline{ac})$.

Drain pipes or trenches placed along a horizontal line in an unconfined aquifer are called *galleries.* Galleries were used for water supply even in prehistoric times. Today these methods are preferred in dewatering (French drains) and in agriculture in order to lower the ground water level. Based on the theories of Dupuit, the discharge from a unit length of a gallery laying on the impermeable bottom of a horizontal aquifer may be derived as

$$\frac{Q}{kHS'} = \frac{2 - (S'/H)}{\ln R} \qquad (6.16)$$

in which Q is the flow from both sides and R is the radius of influence, the distance at which the drawdown is not noticeable. Information in Figure 6.9 aids in the solution of this equation.

Horizontal collector wells are similar to galleries arranged in a radial manner, where the drain pipes of L length discharge water into a sealed concrete shaft. The shaft is first sunk into the ground with valved stubs of pipes placed into the shaft's wall before sinking it. The collector pipes are then pushed into the soil horizontally by using hydraulic jacks. Construction of a horizontal collector well is considerably more expensive than drilling vertical wells, but their significantly greater yield more than offsets the difference in construction costs, particularly if the aquifer is relatively shallow. The yield of a single horizontal collector well is often greater than seven to 15 conventional wells in a well field at the same location. The discharge formula for horizontal collector wells was determined by laboratory experiments using sand models to be

$$\frac{Q}{kLS} = 1.2 + \frac{0.9}{n} \qquad (6.17)$$

where n is the number of symmetrically placed collectors each of length L. Other variables are shown in Figure 6.10. In the experiments on which Equation 6.17 is based the relative head H/L ranged between 0.5 and 1.4. Field data collected in Europe and the United States indicates that considerable lateral variation of permeability is encountered with horizontal collector wells. This in turn results in significant deviations from the design discharge obtained by Equation 6.17. The Q/kLS value was found to range between 0.08 and 2.5 in actual practice, while in our laboratory experience with sand model, where uniform permeability was maintained, it ranged between 0.7 and 1.2. The number of collector pipes found in actual installations ranged between two and 12. To obtain the expected discharge from a collector well the number of collector pipes can be increased almost at will.

EXAMPLE 6.9

A horizontal collector well is built with 8 collector pipes each 50 ft long. The static head of the aquifer is 18 ft. The assumed average permeability of the aquifer composed of coarse, clean sand is 10^{-1} cm/s. Estimate the probable discharge of the well, when the drawdown in the shaft is 8 ft.

SOLUTION:

By using Equation 6.17 with

$$k = 10^{-1} \frac{cm}{s} = 3.28 \times 10^{-3} \frac{ft}{sec}$$

$$L = 50 \, ft$$

and

$$S = 8 \, ft$$

we get

$$Q = \left(1.2 + \frac{0.9}{n}\right) k \cdot L \cdot S$$

$$= \left(1.2 + \frac{0.9}{8}\right) 3.28 \times 10^{-3} \times 50 \times 8$$

$$Q = 1.72 = 772 \text{ gpm}$$

Since $H/L = 18/50 = 0.36$, the problem is not within the proven range of validity of Equation 6.17 but still furnishes a good estimate.

Artesian wells obtain their water from one or more water-bearing layers located at greater depths in which the water is under hydrostatic pressure. For the discharge of a horizontal artesian layer of T thickness, Dupuit's derivation yields

$$\frac{Q}{kTS} = \frac{2\pi}{\ln (R/r)} \qquad (6.18)$$

where R, again, is the radius of influence given by Equation 6.14.

EXCEPTIONAL SEEPAGE PROBLEMS

The physical concepts used in common applications of seepage theory are not necessarily valid in all cases encountered in nature. For example, the problems involving free surface flow are generally analyzed under the assumption that the free surface is the upper limit of the flow field. In soils containing silt and clay a *capillary layer* develops. This capillary layer can sometimes be thicker than the flow field proper. Table 6-2 shows the capillary rise that is encountered in some soils. The actual flow field in soils conducive to capillary action is much larger than that suggested by the classical theory of ground water flow. Although in the capillary fringe the water pressure is negative, it is no barrier to the flow. These problems are treated by recognizing that the capillary layer, or at least most parts of it, is a part of the whole flow field. When a capillary layer is present on the top of a gravity flow field the free surface will not

Table 6-2 Capillary Rise in Various Soils

Soil Type	Average Grain Size in Millimeters	Capillary Rise in Meters
Sand	2–0.5	0.03–0.1
	0.5–0.2	0.1–0.3
	0.2–0.1	0.3–1.0
Silt	0.1–0.05	
	0.05–0.02	1.0–3.0
Silt clay	0.02–0.006	3–10
	0.006–0.002	10–30
Clay	< 0.002	30–300

act as a boundary to the flow. It will not be a limiting stream surface, only a piezometric surface, representing the location of atmospheric pressure in the flow field; hence, the elevation to which water will rise in an exploratory bore hole (Figure 6.14).

Since the capillary layer thickness depends on the surface tension, adhesion, and pore size, this layer will be nearly parallel to the piezometric surface and in case of gravity flow in the flow field proper, the water will move within the capillary layer as well. Since on the upstream side of the flow field the outside water level is below the capillary layer, the water flowing in the capillary layer must enter through the piezometric surface along its upper portion (Figure 6.14). To facilitate this movement there must be an energy gradient upward through the upper entry portion of the piezometric surface. This necessitates the lowering of

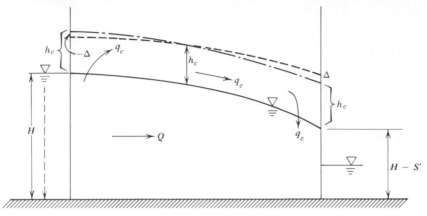

FIGURE 6.14 The mechanism of capillary flow over a piezometric surface bounding a gravity flow field.

the capillary height within the entry portion of the capillary layer. From here the water in the capillary layer will flow along with the rest of the moving water toward the downstream portion of the capillary layer. Because of the negative pressures within the capillary layer the water cannot exit from this layer directly to the atmosphere but must leave the capillary layer through the downstream portion of the piezometric surface. Since this requires excess pressure along the exit portion, the capillary height must increase there. Therefore, the top of the capillary layer over a gravity flow field will always have a flatter slope than the piezometric surface in the flow field.

Experimental investigations of the flow within a capillary layer on the top of a gravity flow field indicated that the velocity varies with a well-defined regularity from zero at the top of the capillary layer to an approximately uniform velocity near the piezometric surface. Considering Darcy's equation (Equation 6.2), one must conclude that since the horizontal energy gradient is not zero at the top of the capillary fringe the permeability must be zero. Indeed, this agrees with the well-known fact that the degree of saturation within the capillary layer decreases with height and only the lower portion is fully saturated. The layer of partial saturation is a layer of varying permeability. The reason for this rests on the great variation in the size of pores in natural soils. Larger pores allow lower capillary heights as it was proved in Chapter 1, and water will rise higher in the smaller pores. Hence the capillary rise in natural soils follows the random function of the pore size distribution. This fact prevents the application of the classical theoretical methods of seepage analysis. But based on several independent studies with sand models, an empirical formula may be used for estimating the discharge q_c of a capillary layer. This may be written as

$$\frac{q_c}{Q} = C \frac{h_c}{H} \left(\frac{S'}{H}\right)^{2.5} \qquad (6.19)$$

in which Q is the discharge of the gravity flow field, H is the static head upstream, h_c is the thickness of the capillary layer and S' is the drop of the piezometric surface across the flow field. The coefficient C in this equation was found to be 0.545 in the sand model experiments.

EXAMPLE 6.10

For the problem discussed in Example 6.7 estimate the probable capillary discharge.

SOLUTION:

The upstream head in the problem was 20 m and the exit point of the free surface measured from the downstream water level was 4 m. In Equation 6.19 used to determine the capillary discharge, the variable S' is to be measured from the upstream water level. Hence, $S' = 20 - (4 + 1) = 15$ m. The discharge Q for unit length of wall was 6.4×10^{-5} m³/s.

The permeability of the soil was 10^{-3} cm/s. From Table 6-1 this indicates a fine sand with an average grain size of 0.1 mm. In Table 6-2 the capillary rise h_c for such soils range between 0.3 and 1.0 m. Taking the maximum value we may substitute into Equation 6.19 as

$$q_c = 0.545 \left(\frac{1}{20}\right)\left(\frac{15}{20}\right)^{2.5} \times 6.4 \times 10^{-5}$$

$$= 8.5 \times 10^{-7} \frac{m^3}{s} \text{ for one meter of wall}$$

Since the wall is 50 m long the total capillary discharge is

$$50 \times 8.5 \times 10^{-7} = 4.25 \times 10^{-5} \frac{m^3}{s}$$

which equals 969 gallons per day.

Another case when the formal theory of seepage breaks down occurs in connection with *seepage losses* from canals and rivers. Some of these problems can be analyzed by regular means, particularly when the ground water level is only a little lower than the level of the water in the channel. In many places, on the other hand, this is not true. When the water level in a porous ground is well below the bottom of the channel, the water in the channel is not connected in a continuous manner with the ground water. Particularly when the channel carries suspended sediments, silt, and clay, it plugs the pores of the channel bottom, creating a relatively less permeable layer on the bottom of the channel. This occurrence is referred to as colmatation. The water that seeps through this layer because of the pressure on the channel bottom enters a more permeable soil after passing through. As the permeability increases downward the escaping water is allowed to flow faster. When this is the case, part of the pores in the soil contain air and as a result the soil is unsaturated. The unsatu-rated soil pores will be in contact with the atmosphere. As such, the pressure there will be atmospheric. The water seeping downward in such cases will behave somewhat like water flowing from a shower. Just as with the elevation of the bathtub, the location of the permanent ground water table below the river will be irrelevant. In these cases the rate of seepage loss will be strictly dependent on the average depth of water, the permeability of the channel bottom layer, and its thickness. The permeability of soil itself, as long as it is larger than that of the bottom layer, will not influence the seepage losses.

UNSTEADY SEEPAGE

All the foregoing considerations of seepage pertain to steady flow. For steady flow conditions a constant supply of ground water discharge is needed. While we know that most of the ground water comes from rainfall, part of which in-filtrates into the ground, the seasonal fluctuations of the ground water inflow affects our assumptions of steady seepage in a negligible manner. In water supply from wells, on the other hand, the ground water inflow into the layers over long periods of time may be significantly less than the amount pumped out. In these cases the yield decreases slowly because of the lack of equilibrium. Only experience over a long period of time can answer such questions.

In the inflow and outflow boundaries of flow fields the sudden variation of the boundary potential because of rapid decrease of controlling water levels can create very large seepage potential differences in the soil. The resulting seepage forces often exceed the internal strength of the soil, causing *slope failures* along the embankments. Shore lines of creeks and rivers often exhibit this phenomenon after floods. Recently this effect caused the catastrophic failure of a dam in Europe. The flood wave caused by a landslide has wiped out a whole village downstream.

Flow through levees that are becoming slowly saturated during floods is another example of unsteady seepage. As compared to seepage velocities the rise of the flood water is relatively sudden. Levees are made of local soil that is usually relatively permeable. Hence they will retard the flow for only a short period of time, until the advancing free surface reaches the protected side. The width of levees therefore is selected so that they hold the water as long as the flood is expected to last. Prolonged floods may cause levees to leak. The danger of failure hence increases with time as leakage, piping, and increasing seepage forces develop.

PROBLEMS

6.1   Water seeps into a trench horizontally from a layer of silty sand that has a permeability of 10^{-5} cm/s. The discharge velocity is estimated to be 10^{-6} cm/s. Express the magnitude of the horizontal seepage force acting on a unit volume of the soil.

6.2   For unit volume of sand in Darcy's sand filter arrangement, shown in Figure 6.1, express the hydraulic energy transferred due to seepage.

6.3   For a permeable soil layer under an earth dam built of clay the form factor was determined to be $A/L = 0.07$ m, in which A is the average cross-sectional area of the permeable soil seam measured perpendicular to the flow and L is the average length of the flow lines. Estimate the seepage discharge for a head loss across the layer if $S = 7$ m and if the assumed permeability is 10^{-3} cm/s.

6.4   For a 50-ft wide concrete dam built on the top of a 100-ft deep horizontal uniform fine sand layer a sheet pile should be driven at the center line to cut the seepage discharge by 30%. How deep should the sheet pile be? What would be the additional reduction of discharge if the sheet pile is driven at the front of the dam?

6.5   A trench built in a 6-ft deep lake involves an 18-ft wide excavation. The soil is silty sand with an average grain size of 0.02 mm. The depth of the excavation is to be 10 ft below the lake bottom. To what depth should sheet piles be driven to result in a maximum exit pressure gradient of 0.5. How much is the seepage discharge in this case, per unit length of trench?

6.6   In the case of Problem 6.5 what would be the hydrostatic pressure on the bottom of the excavation if a 1 ft thick concrete mat would be poured to eliminate the flow of the water?

6.7 A rock filled dam built with a 20-ft thick vertical clay core is resting on an impermeable bedrock. The permeability of the clay is 10^{-6} cm/s. Determine the seepage discharge for a 100 ft length of the dam for the case when the upstream water is 20 ft above the base and the downstream water level is zero. How high would be the zone of saturation (surface of seepage) on the protected side, neglecting capillary effects?

6.8 A retaining wall built to support a saturated 4 m tall sand layer ($k = 10^{-2}$ cm/s) is provided with a vertical drain of well-graded sand and gravel resulting in 100% drainage on the wall. How much would be the height of the surface of seepage in the sand behind the drain if the discharge per unit length is 0.00016 m^3/s?

6.9 Assuming that the permeability of a 10-ft thick shallow horizontal aquifer is $k = 10^{-2}$ cm/s, determine the discharge of a horizontal collector well built with 6 symmetrically placed 30-ft long collector pipes, if the drawdown is 6 ft in the shaft. How many conventional 8 in. diameter drilled wells would have to be built and operated with the same drawdown to provide an equivalent amount of flow?

6.10 Determine the capillary discharge in Problem 6.7 if the height of the capillary layer is 15 ft.

ELEMENTS OF HYDROLOGY

In the design and analysis of open channels and hydraulic structures the most important variable is the magnitude of the flows that may be expected. Yet this variable is the least certain of all the variables since it depends on many interrelated topographic, geologic, and other elements of nature, in addition to its dependence on random meteorological factors. Fundamental principles and methods used in the determination of design discharges are reviewed in this chapter.

Hydrology is the study of the circulation of water in Nature. It covers the physics of what is called the *hydrologic cycle*, from the evaporation of the seas through the movement of the atmospheric moisture that makes up the weather, the statistics of rainfalls, to the collection and runoff of rainwater through creeks and rivers. Our particular interest in this subject is focused on the magnitude and occurrence of floods in order to determine reasonable values for design discharges pertaining to open channels and hydraulic structures.

THE DESIGN DISCHARGE

In hydraulic design few things are more important than the selection of the design discharge. This is the maximum capacity a channel or structure is designed to handle.

Basically the matter is a question of economics. The smaller the design discharge of a structure the cheaper it is to build. The most important question is: What will happen if the design discharge is exceeded? In the case of a highway culvert we may be willing to put up with the occasional flooding of a few acres of agricultural land. Even the threat of periodic payment of damages would not deter the designer from selecting a design discharge corresponding to a rainfall expected once in five years or so. Designing his culverts to the size corresponding to a hundred years flood would result in far too big an investment. On the other hand, an undersized spillway of a dam presents a great deal more danger. A dam may fail once its crest is overtopped, causing considerable losses in property and in the even more critical cases, life as well.

The period within which we cannot

tolerate the design discharge to be exceeded by the random flows due to rains is called the *return period* of a flood. Recurrence interval or flood frequency are other expressions used for the same thing. The return period of a flood of certain magnitude is a statistical concept. The larger the flood, the less frequently it will be encountered. As all matters of random nature analyzed by statistical methods, flood frequencies represent a degree of uncertainty. The hapless designer selecting, for example, a 10 year design discharge could find his structure flooded out twice in the first two years, once by a 11 year flood, then by a 500 year one. Selecting a return period that exceeds the expected life of a structure is no assurance that larger flows will not be encountered within the life of the structure. Their occurrence is a matter of chance, which we try to counter with our educated guesses. How educated our guesses are may later be determined in the courtroom. On the one hand, a designer will not stay in

business long if his designs are consistently more expensive than those of his competitors. However, the designer should refrain from wishful thinking.

RAINFALL STATISTICS

Floods are generally caused by rains. Those caused by melting snow are only a more complicated variety. Rainfalls are measured by meteorological stations located at most airports and other sites scattered about the country. Data on rainfalls going back sometimes over a hundred years or more at some locations were analyzed by statistical means. Such data is represented by three important variables. These are the intensity, the duration, and the frequency of rainfall.

Intensity is computed by dividing the rainfall depth of a particular rain with the period of time during which the rain has fallen. It is expressed in inches or centimeters per hour. Statistical findings have

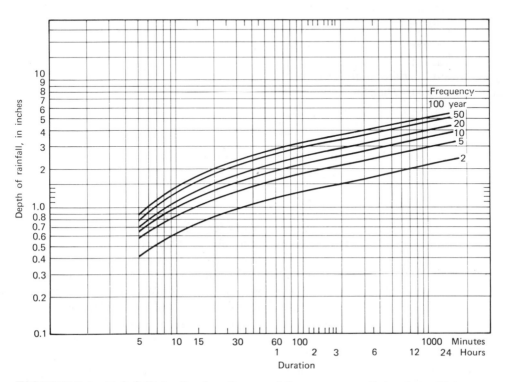

FIGURE 7.1 Rainfall depth, duration, and frequency at Columbus, Ohio.

proven that generally the shorter the time during which a rain is falling, the higher its intensity.

Duration is the period of time during which a particular rain is falling. It is expressed in minutes, hours, or days.

Frequency of a rain refers to the return period of a particular rainfall characterized by a given duration, intensity, or both. It is expressed in years. Because of the relatively short period of time during which rainfalls were measured at most places, statistical data is extrapolated on the basis of probability theory. This allows us to estimate rainfall and flood occurrences up to frequencies of 100 to 500 years even though data is available for a much shorter period. Statistical data of this type is more reliable than the raw data since the information collected directly has a random nature. As such, a period of 30 years of measurement could easily contain rainfalls corresponding to frequencies much greater than 30 years. In 30 years of rainfall measurements we may well have encountered several rains of 50, 100, or even 500 year frequencies. These become obvious only after the data was subjected to statistical analysis.

Statistical analysis allows us to combine rainfall or flood data measured at several points in the geographic region in question. The method used for lumping these various sources of information is called multicorrelation. Because of the very large amount of data to be considered in this case, the analysis is done by computers. As an example of these results, rainfall depth, duration, and frequency curves developed for Columbus, Ohio using multicorrelation are shown in Figure 7.1.

EXAMPLE 7.1

Determine the return period and intensity of a rainfall in the Columbus, Ohio area for a rain lasting two hours if the total rainfall depth was measured to be 3 in.

SOLUTION:

By definition the rainfall intensity is the rainfall depth divided by its duration in hours; hence,

$$I = \frac{3}{2} = 1.5 \frac{\text{in.}}{\text{hour}}$$

From Figure 7.1 a 2 hours duration rain at 3 in. rainfall depth corresponds to a 50 year return period.

Rainfall data for a particular geographic area is generally available for the designer of hydraulic projects from federal, state, or local sources. The U.S. Weather Service of the National Oceanographic and Atmospheric Administration, State Departments of Natural Resources, State Highway Departments, nearby universities and research organizations should be contacted for hydrologic data in an early stage of the work. In developing countries sometimes the best information is available from pertinent organizations of the former colonial powers.

Experience with storms showed that heavy rains are associated with localized storm patterns of approximately elliptic shape that are aligned with the weather fronts. The concentration of rains over such storm patterns means that the smaller the area the larger is the possible maximum precipitation. Rainstorms creating the largest floods in a drainage basin are those whose storm patterns best overlap the drainage basin. If the topographic shape of a drainage basin extends in the general direction of the weather fronts commonly encountered in that region, large floods occur more frequently than in drainage basins that are perpendicular to the frontal patterns. Maximum possible 24 hour precipitations over 200 square mile areas within the United States east of the Rocky Mountains are shown in Figure 7.2. To convert these data to areas of other sizes and for rains of larger dura-

tions, Figure 7.3 may be used. The index rainfall ratio shown in this graph refers to the maximum possible rainfall as a percentage of the 200 square mile, 24 hour rainfall obtained from Figure 7.2. Projects are rarely designed to handle the maximum possible precipitation unless the maximum possible safety is required, as in the design of spillways for high dams. Experience of the United States Army Corps of Engineers indicate that about 40 to 60% of the maximum possible precipitation is usually selected as a basis of design for flood control projects.

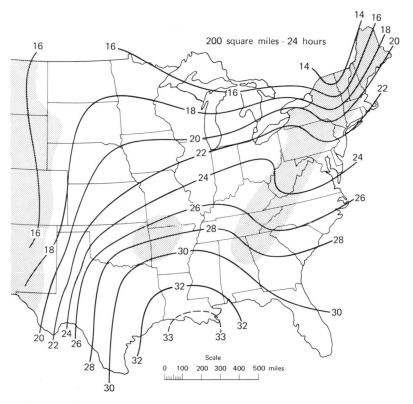

FIGURE 7.2   Maximum possible precipitation in inches over a 200 square mile area in one day.

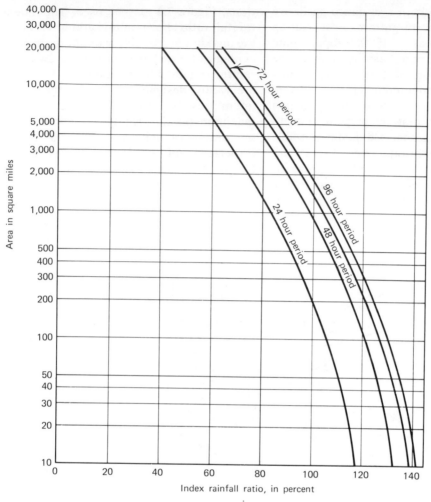

FIGURE 7.3   Depth to area to duration relationships from storm studies by the U.S. Army Corps of Engineers.

EXAMPLE 7.2

Estimate the maximum possible 48 hour rainfall over a 20 square mile area at the vicinity of Charleston, West Virginia.

SOLUTION:

The maximum possible rainfall at this location for 24 hours of rainfall over a 200 square mile area is indicated in the map of Figure 7.2 to be 24 in. By using Figure 7.3 we find that the 48 hour rainfall for a 20 square mile area is 130% of the 24 in., or 31.2 in.

THE MAGNITUDE OF FLOODS

Rainfall is only one of several parameters that contribute to the magnitude of floods encountered at a particular site. Other important parameters include the size of the land area from where the rainfall is collected, the shape of this area, and its average slope along the main channel through which the rainwater is led to the site in consideration. The type of soil, whether it is permeable or impermeable, cultivation, land use, and similar factors all contribute to the relative magnitude of a flood. Of all these variables the most important one for flood magnitude is the land area from where all rains collect to the site of the structure designed. This is called the *watershed* of the particular outflow point. The size of this drainage area is to be measured from topographic maps by the designer. Distinction is made by experts of hydrology between small watersheds and large watersheds. Drainage areas of less than 30 square miles (about 80 km^2) are considered small watersheds. For small watersheds the most important factor after the size of drainage area is the rainfall intensity. Second to this is the slope of the main channel. For large watersheds the most important variable after the drainage area size is the main channel slope, followed by the rainfall intensity.

Once the rain has started to fall it will begin its pattern of flow along the steepest descent of the land. The rate of collection depends on the relative density of available flow channels. Water flows faster in creeks and other channels than on land. The velocity of the flow depends on the slope of the channels. For hydrologic computations the term "main channel slope," S, is the ratio of the difference in elevation between the points 10 and 85% upstream from the studied site to the watershed boundary divided by the length of channel between these points. Hence the steepest 15% and the flattest 10% of the land contours are excluded in this consideration. The main channel slope is determined by the design on the basis of available topographic maps.

The time it takes for the first raindrop fallen at the most distant point of the drainage area to reach the outlet of the watershed is called the *time of concentration, T_c*. Experience proved that the most critical floods are of rains whose duration at least equals the time of concentration. For small drainage basins of agricultural land the time of concentration in minutes may be estimated from the formula

$$T_c = 0.0078K^{0.77} \qquad (7.1a)$$

in which K equals the maximum length of travel in feet divided by the square root of the main channel slope S. Another formula recommended for use in conjunction with drainage of airports is

$$T_c = \frac{1.8(1.1 - c)\sqrt{D}}{\sqrt[3]{S}} \qquad (7.1b)$$

in which c, the soil index, a somewhat uncertain parameter that is less than 0.4 for pervious ground, and ranges between 0.3 and 0.65 for impervious soils and between 0.7 and 0.95 for pavements and roof surfaces. Of the other variables, D is the longest drainage distance and S is the mean slope.

The effect of surface soils and geologic characteristics is known to be highly important to flood peaks. Depending on the permeability of the land, part of the rainfall during its initial period infiltrates into the ground. Rainfall up to 0.5 in. may be absorbed into the soil in humid regions, provided the soil was relatively dry prior to the beginning of the rainfall. The rate of infiltration depends on the soil cover, land use, and cultivation. Soil infiltration rates, S_i, in terms of inches per hour are often available from state conservation services, U.S. Department of Agriculture, and similar organizations.

Flood discharges for small and large watersheds were measured on a regular basis at many locations. The correlation of measured flood magnitudes with the corresponding rainfalls, drainage areas, main channel slopes, soil infiltration coefficients, and other pertinent variables generally result in an equation of this type:

$$Q_T = aA^bS^cI^dS_i^e \qquad (7.2)$$

Here Q_T in ft³/sec is the peak discharge of a flood with a given recurrence interval T. The lower case letters "a" through "e" are constants determined by correlation and valid for particular geographic regions; A is the drainage area in miles²; S is the main channel slope expressed in ft/miles; I is the rainfall intensity in in./hour corresponding to the recurrence interval T sought for the flood Q_T; and S_i is the soil infiltration in in./hour. In equations of the type of Equation 7.2 the duration of the critical rainfall is usually taken to be the 24 hour intensity rainfall with a $T = 2.33$ years duration. An example for Equation 7.2 is given in Table 7-1 showing the values of the coefficients computed on the basis of a large-scale multicorrelation of data collected in Northeastern Ohio. Similar data may be found in hydrologic literature for other areas.

Table 7-1 Flood Discharge Formulas for Northeastern Ohio

Drainage Basin Type	T, in years	a	b	c	d	e
Small watershed	2.33	1.9	0.84	0.15	4.03	0.43
(less than	10	3.2	0.81	0.14	4.14	0.41
40 miles²)	20	4.6	0.80	0.15	4.00	0.43
	50	7.3	0.79	0.15	3.70	0.40
Large watershed	2.33	5.2	0.86	0.39	1.62	− 0.15
	10	12.0	0.82	0.36	1.58	− 0.15
	20	12.2	0.82	0.39	1.65	− 0.15
	50	16.5	0.80	0.35	1.72	− 0.17

$$Q_T = aA^bS^cI^dS_i^e$$

Note: I is the intensity in inches per day with a constant recurrence interval of 2.33 years.

EXAMPLE 7.3

If a 25 square mile drainage area north of Columbus, Ohio has a mean channel slope of 2% and the soil infiltration is 6 in./hour determine the expected maximum flood that has a recurrence interval of 10 years.

SOLUTION:

From Equation 7.2

$$Q_T = aA^bS^cI^dS_i^e$$

in which

$$A = 25 \text{ square mile}$$

$$S = 0.02 \text{ ft/ft} = 0.02 \times 5280 = 105.6 \frac{\text{ft}}{\text{mile}}$$

From Figure 7.1

$I = 2.5$ in./24 hours at 2.33 years frequency,

$S_i = 6$ in./hour

$T = 10$ years

From Table 7-1 $a = 3.2$, $b = 0.81$, $c = 0.14$, $d = 4.14$, $e = 0.41$
Then

$$Q_{10} = 3.2(25)^{0.81}(105.6)^{0.14}(2.5)^{4.14}(6.0)^{0.41}$$

$$= 3.2(13.56)(1.92)(44.4)(2.08)$$

$$Q_{10} = 7694 \text{ cfs}$$

When data of the type shown in Table 7-1 is unavailable, flood magnitudes may be estimated by what is known as the *Rational Formula*.

The rational method assumes a direct relationship between discharge, critical rainfall intensity, and drainage area. It is expressed by the equation

$$Q_T = cI_TA \qquad (7.3)$$

in which Q_T (acre-in./hour)* is the flood discharge for a recurrence interval T, I_T is the rainfall intensity in in./hour for a duration corresponding to the time of concentration T_c obtained from Equation 7.1 or by other means. A is the area of the drainage basin in acres. The coefficient c in Equation 7.3 depends on the land surface and is obtained from Table 7-2.

The range of values given in Table 7-2 indicates the degree of uncertainty in the rational method. But because of its simplicity its use is rather popular for the design of small hydraulic structures, particularly in urban areas.

Table 7-2 Coefficients for use with the Rational Formula (Equation 7.3)

Description of Area	Runoff Coefficients
Business	
Downtown	0.70 to 0.95
Neighborhood	0.50 to 0.70
Residential	
Single-family	0.30 to 0.50
Multi-units, detached	0.40 to 0.60
Multi-units, attached	0.60 to 0.75
Residential (suburban)	0.25 to 0.40
Apartments	0.50 to 0.70
Industrial	
Light	0.50 to 0.80
Heavy	0.60 to 0.90
Parks, cemeteries	0.10 to 0.25
Playgrounds	0.20 to 0.35
Railroad yard	0.20 to 0.35
Unimproved	0.10 to 0.30

* One acre-in./hour nearly equals 1 ft³/sec.

EXAMPLE 7.4

A 30 acre single family residential development is built on gently sloping land. The general slope of the land is 6 ft/mile. The maximum length of travel for the runoff water is 1500 ft. Estimate the 5 year design discharge for a drainage channel, assuming that the local rainfall statistics correspond to the data on Figure 7.1.

SOLUTION:

The time of concentration T_c is computed by Equation 7.1 as follows:

$$S = \frac{6 \text{ ft}}{1 \text{ mile}} = \frac{6}{5280} = 0.0011 \frac{\text{ft}}{\text{ft}}$$

$$T_c = 0.0078 \left(\frac{1500}{\sqrt{0.0011}}\right)^{0.77} = 30 \text{ min}$$

From Figure 7.1 the 30 min duration rain at 5 year frequency corresponds to a rainfall depth of 1.49 in. of rain. The intensity of this rain is $I_s = 2.98$ in./hour.

By using Equation 7.3, we get

$$Q_T = c I_T A$$

With $c = 0.35$ from Table 7.2 we get

$$Q_s = 0.35 \times 2.98 \times 30 = 31.29 \text{ (acre in./hour)}$$

Using the conversion table in Appendix II we find that this discharge equals

$$Q = \frac{31.29}{12} \times 24 = 62.58 \text{ acre } \frac{\text{ft}}{\text{day}}$$

$$= 62.58 \times 0.504 = 31.54 \text{ cfs}$$

(Since 1 acre-ft equals 1.008 cfs.)

There are several other means to obtain the magnitude of flood discharges pertaining to a selected recurrence interval. These include semigraphical techniques, such as the Unit Hydrograph method, the Standard Project Storm method used by the U.S. Army Corps of Engineers, which is based on storm patterns found to give critical floods in certain geographic areas, the Chicago Hydrograph method used for sewerage design, and others. Information on these are available in the technical literature.

FLOOD ROUTING

Once the flood waters of a rain are flowing in a channel the flood wave soon passes the area over which the rain has fallen. The analysis of the flow of a flood wave in a channel is called flood routing. Routing of a flood means the determina-tion of the variation of the discharge at the downstream end of a portion of a certain length of channel, called a reach, if the upstream discharge variation is known.

At the edge of a watershed over which the rain has fallen, the maximum, peak discharge can be computed by the methods described above. Knowing the discharge before the rain, the so-called base discharge, and knowing the approximate value of the time of concentration we may construct a curve showing the approximate relationship between the discharge and time. A nearly straight S-shaped curve between the base discharge Q_B and the peak discharge Q_p over the time T_c will make a reasonable approximation when field gage measurements are lacking. Figure 7.4 shows such a relationship. This is called inflow hydrograph of the reach below. The portion of the curve beyond T_c was drawn as the mirror image of the first part of the curve.

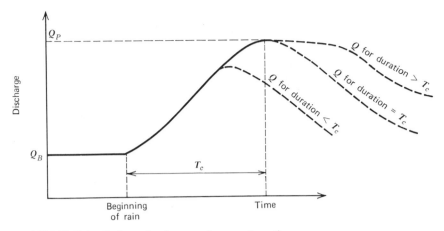

FIGURE 7.4   Inflow hydrograph construction.

The reach may include a reservoir below which a hydraulic structure is to be analyzed. Alternately the reach includes a portion of a river followed by subsequent reaches. The methods to be outlined work for both cases, giving the outflow hydrograph for a reach on the basis of the known inflow hydrograph. The outflow hydrograph becomes the inflow hydrograph of the subsequent reach downstream.

Knowing the length, surface area, and the shape of the reservoir or the general channel section of the reach in question, we can construct a graph showing the depth of the water versus the storage in the channel as shown in Figure 7.5. By continuity we know that for any given time period Δt the difference between the flow entering and leaving the reach must equal the change of storage in the reach. Denoting the flow at the beginning of Δt as subscript 1 and at the end of Δt as subscript 2 of the inflow and outflow discharges, we can write our continuity relationship as

$$\frac{I_1 + I_2}{2} \Delta t - \frac{O_1 - O_2}{2} \Delta t = \Delta S = S_2 - S_1$$

$$(7.4)$$

in which ΔS refers to the change in storage within the reach during the period of Δt.

The initial value of the outflow dis-

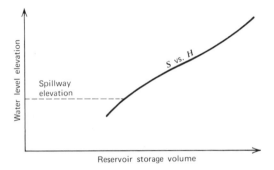

FIGURE 7.5   Storage as a function of depth.

charge O_1 is known to be equal to I_1; both are equal since the base flow is constant before the rain. Also, the initial value of the storage is known. Furthermore I_1, I_2, O_1, S_1 in Equation 7.4 are initially known, so is Δt as this may be selected by us to be a convenient time period, an even fraction of T_c.

Rearranging Equation 7.4 to group the unknown values to the right side, we obtain

$$I_1 + I_2 + \frac{2S_1}{\Delta t} - O_1 = \frac{2S_2}{\Delta t} + O_2 \quad (7.5)$$

This gives us one equation with two unknowns, too many to solve the problem directly.

In the case of reservoirs where for the sake of spillway design a critical discharge is sought, there is one additional relationship. For a spillway the discharge may be determined as a function of the height of the water over the spillway crest (see Equation 9.9). The relationship is shown by the so-called *rating curve* of the spillway, an example of which is shown in Figure 7.6. The water elevations shown in

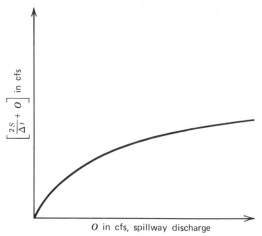

FIGURE 7.7 Combined reservoir storage spillway outflow graph to solve routing equation.

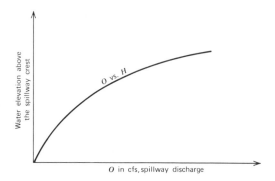

FIGURE 7.6 Spillway rating curve.

Figures 7.5 and 7.6 are of course identical. Hence, the two graphs give a relationship between the two unknown variables S and O in Equation 7.5. For convenience in computations a new graph may be constructed using Figures 7.5 and 7.6, selecting a few water elevations to form corres-

ponding values of $(2S/\Delta t) + O$ as shown in Figure 7.7. The graph shown in Figure 7.7 now simplifies Equation 7.5 by making its right side a function of the outflow only.

The outflow hydrograph, that is, the value of O_2 for all subsequent time periods of Δt, can now be computed in a step by step manner, using the value of O_2 as O_1 for the next time period.

EXAMPLE 7.5

Floods from a 5.87 square mile drainage area enters a reservoir which has a 2170 ft² surface area bounded by steep shores such that the surface area is essentially constant for small variations of the water level. The water leaves the reservoir through a weir that has a width of 24 ft. From hydrologic analysis it was determined that the 50 year flood results in a peak discharge of 855 cfs. The time of concentration was computed to be 100 min. Route this flood through the reservoir under the assumption that the reservoir is full to the crest of the weir at the beginning of the rain, and that the duration of the rain is 100 min.

SOLUTION:

For the sake of simplicity the inflow hydrograph is assumed to be a triangle with its apex at Q_{max} equals 855 cfs at 100 min from the beginning, the triangle being symmetrical to its apex. (Other, bell-shaped inflow hydrograph could have been drawn also, resulting in some improvement in the results.) Therefore, the discharge values that increase linearly with 10 min time intervals are shown in the table of computations below. Column 1 shows the successive time intervals, Column 2 shows the corresponding inflow discharges. From Columns 1 and 2 the volume of water entering the reservoir is computed based on the averaged discharge value for each time interval; this is shown in Column 3.

The outflow from the reservoir is controlled by the weir flow equation (Equation 9.9), which in this case was determined to be

$$Q_{out}(cfs) = 1.65\sqrt{g}(width)(h^{3/2}) = 224.7(h^{3/2})$$

where h (ft) is the height of water in the reservoir over the crest of the weir. To facilitate the computations a graph may be constructed to show the relationship above. (For a similar graph see Figure 7.6.) Since the surface area of the reservoir is constant for small variations of h, the storage at any time in the reservoir is directly related to the height h. Otherwise a graph similar to the one shown in Figure 7.5 must be constructed from topographic studies.

At the beginning of the flood the initial storage, head, and hence the outflow discharge is zero. For the end of the first time period (10 min) the outflow discharge is so small that it may be assumed to be zero also. For each subsequent time period the added storage

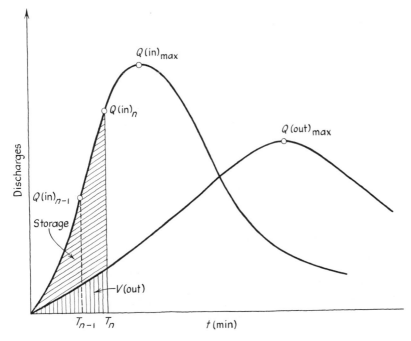

FIGURE E7.5 V_{in} = storage + V_{out}. $T_n - T_{n-1} = \Delta t$.

may be computed by using Equation 2.2a. After determining the first rise in the reservoir level the outflow discharge is determined from the weir flow equation or its graphical representation. The outflow discharge is shown in Column 7. After computing the first outflow discharge value the volume of water remaining in storage can be determined and is shown in Column 5. Figure E7.5 explains the basis of these computations. The calculation then proceeds step by step for each subsequent time period. After carrying out the computations shown in the table we find that the maximum outflow discharge on the weir is 487 cfs. This peak outflow occurs 160 min after the rain starts and 60 min after it stops. For the case when the duration of the rain exceeds the time of concentration according to our hydrologic studies (e.g., Figure 7.1), the inflow hydrograph should be adjusted by extending the peak flow horizontally to the inflow hydrograph. This would result in a higher outflow value, ultimately stabilizing at the level of the peak inflow.

For the situation shown in the table the reservoir reduces the outflow by 57%.

Table of Computations for Example 7.5

1	2	3	4	5	6	7
T Time (min)	Q (in) (cfs)	V (in) (ft^3)	V (out) (ft^3)	S (Storage, ft^3)	h (Height, ft)	Q (out) (cfs)
0	0	0	0	0	0	0
10	10	3	0	3	0.0014	0.0116
20	70	27	0.0035	26.99	0.0124	0.309
30	200	108	0.1	107.9	0.050	2.49
40	370	279	0.9	278.1	0.128	10.27
50	510	543	4.8	538.2	0.248	27.67
60	620	882	16.2	865.8	0.399	56.5
70	710	1281	41	1240	0.571	96.7
80	775	1726	78	1639	0.755	147.0
90	820	2205	161	2044	0.94	205
100	835*	2701	265	2436	1.12	266
110	810	3195	407	2788	1.28	326
120	780	3672	585	3087	1.42	379
130	720	4122	796	3326	1.53	425
140	660	4536	1037	3499	1.61	458
150	580	4908	1302	3606	1.66	479
160	470	5223	1584	3639	1.68	487*
170	350	5469	1876	3592	1.66	479
180	230	5643	2166	3477	1.60	454
190	220	5778	2446	3332	1.54	428
200	210	5907	2710	3197	1.47	399

Columns 3, 4, and 5 are in terms of 1000 ft^3.
* Starred discharges represent peak flows.

The flood routing computation for rivers with no control structures (spillways or weirs) is a more involved procedure, because the problem is of great mathematical complexity. For more information on this subject, the interested reader is referred to the literature.

For rough computations one may apply the technique used by hydraulic engineers at the U.S. Army Corps of Engineers. This technique requires that the designer have at least one set of field data on a flood in the reach in question. Particularly the time difference between flood peaking at the inflow and outflow sections must be known. With this time lag K, the relationship between the storage, and the inflow and outflow values can be written as

$$S = K[xI + (1 - x)O] \qquad (7.6)$$

This states that the storage is a function of the inflow as well as the outflow, according to a weighing coefficient x. The value of this weighing coefficient is usually taken to be about 0.35 for rivers in the Appalachian region.

With the assumed K and x values the outflow O_2 at subsequent time periods Δt can be computed step by step using the formula

$$O_2 = c_0 I_2 + c_1 I_1 + c_2 O_1 \qquad (7.7a)$$

where the c coefficients are calculated for the reach before the computations begin, from the equations

$$c_0 = \frac{-Kx + 0.5\,\Delta t}{K - Kx + 0.5\,\Delta t}$$

$$c_1 = \frac{Kx + 0.5\,\Delta t}{K - Kx + 0.5\,\Delta t} \qquad (7.7b)$$

$$c_2 = \frac{K - Kx - 0.5\,\Delta t}{K - Kx + 0.5\,\Delta t}$$

(Note that $c_0 + c_1 + c_2$ must equal one.) To route a flood for a river using this approximate method, we start at the stream end where all I_1, I_2, I_3—and O_1 values are presumed to be known. Using Equation 7.7a we can compute O_2. For the next steps the subscripts are reduced by

one, making O_2 into O_1 and I_2 into I_1, etc.

The design discharge can be obtained for most hydraulic structures by the techniques outlined in this chapter. Because of the complexity of some of these techniques, such as the use of Equation 7.5 and similar equations, the computations could be somewhat cumbersome if design discharges of a number of return periods are sought. The reader might be reminded that flood discharges of various return periods also follow the laws of statistics as rainfalls do. It is sufficient to determine only two values for Q_T, for example, one for a 5 year return period and one for a 50 year flood. Statistical probability papers are commercially available; most bookstores and office supply stores stock them. Plotting the 5 and 50 year discharges on such a probability paper and connecting these points with a straight line will interpolate our results in a reliable manner.

In precise hydrological statistics more refined probability plots are used to extrapolate data. For rainfalls Gumbel's theory of extreme values (like the chart on Figure 7.1) provides plotting positions. For flood analysis the U.S. Army Corps of Engineers prefer the use of Beard's probability function. Studies of these kinds are beyond the scope of this book.

FLOW-DURATION ANALYSIS

The foregoing computations give us maximum expected flows at a particular site. Between these floods the discharge varies in a random manner. By using the methods of *stochastic analysis*, we may obtain the likely sequence of these expected flows. This, however, is beyond the scope of this book.

Depending on the site's watershed and meteorological characteristics, the duration of time during which any given discharge is exceeded in the course of a year may be plotted as the graph in Figure 7.6.

Graphs of this type are valuable in the analysis of the expected performance of a structure. To construct graphs of such kinds it is necessary that a long series of daily flows be available. For small hydraulic structures this is a rare information to have.

By periodic stage or discharge measurements flow-duration graphs as shown in Figure 7.8 may be developed. As the graphs show this type of information can aid the designer of a structure in the evaluation of the proposed outflow devices for discharges less than the maximum design flood, which may occur in rare instances only. Some parts of the outflow structures may be designed for the average daily flows with provisions for the occasional floods. The determination of such average flows is greatly aided by flow duration graphs.

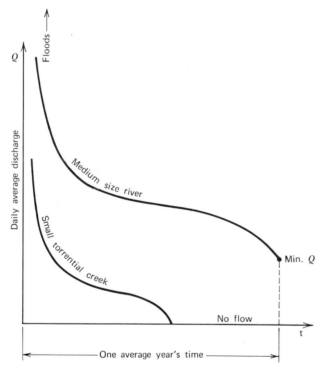

FIGURE 7.8 Discharge-duration graphs for yearly flow of rivers.

PROBLEMS

7.1   Determine the rainfall depth of a 6 hour rain with a 5 year frequency at Columbus, Ohio (Figure 7.1). What is the intensity of this rainfall?

7.2   Determine the maximum possible precipitation at New York, N.Y., that may be encountered on a 15 square mile drainage area over a 72 hour period.

7.3   Determine the time of concentration in an agricultural land where the main channel slope is 0.0048 and the longest drainage distance is 1.3 miles.

7.4   Compute the 50 year flood on a large creek in Northeastern Ohio if the total drainage area is 35 miles^2. The average soil infiltration is 2.5 in./hour, the main channel slope is 0.008. See Figure 7.1 for the assumed rainfall intensity.

7.5 A 16 acre parking lot for which the runoff coefficient may be taken as 0.9 is located near Columbus, Ohio. Determine the five year flood discharge using the "rational formula," assuming that the time of concentration is 10 minutes.

7.6 Based on the 50 year flood discharge computed in Problem 7.4 and assuming a time of concentration of two hours, route the flood through a reservoir with a surface area of 37 acres; the flood is essentially constant for all water levels encountered. The outflow from the reservoir is controlled by a weir. The discharge through this weir is dependent on the height of water over the weir crest, h, according to the formula $Q_{outflow} = 100\sqrt{2g}h^{3/2}$. State the peak flow over the weir for a 3 hours rainfall.

7.7 Route a flood through a reach of a creek, taking into account the storage in the channel, if the base flow is 100 cfs and the incoming flood flow increases 50 cfs for each 2 hours over a 6 hour interval and then subsides by the same rate. From past observations the time lag between peaks at the inlet and outlet of the reach is $2\frac{1}{2}$ hours and x may be assumed to be 0.35.

7.8 By continuous discharge measurements the discharge in a stream was determined over a 3 year period. After assembling the data it was found that the discharge exceeded 120 cfs during 3% of the time, exceeded 60 cfs in 35% of the time, and exceeded 20 cfs in 80% of the time. For the remaining periods the discharge exceeded the minimum 5 cfs flow. Construct a discharge duration graph for the average year. Determine the average discharge in the stream.

3

OPEN CHANNEL FLOW

The concepts relating to flow in channels with a free surface are certainly the most complex ones of the science of hydraulics. Yet their practical relevance is boundless in the understanding of the behavior of water in creeks, rivers, and manmade channels. This chapter is a survey of the fundamental open channel concepts that are essential for the practical hydraulician.

THE ENERGY AT ONE POINT

When flow takes place in a channel or pipe such that the water has a free surface exposed to the atmosphere we speak of open channel flow. Rivers, creeks, sewers, irrigation ditches and drainage channels, culverts, spillways, and similar manmade structures are designed and analyzed by the methods of open channel hydraulics.

The primary difference between confined flow in pipes and open channel flow is that in open channels the cross-sectional area of the flow is not predetermined as in pipes but is a variable that depends on many other parameters of the flow. For this reason hydraulic computations related to open channel flow are the most complicated aspects of the science of hydraulics.

On the other hand, as the pressure on the open surface is always that of the atmospheric pressure the hydraulic grade line in open channels coincides with the water surface. In pipe flow the hydraulic grade line is independent of the position of the pipe; the pressure causes the movement of the water. There is no pressure energy to drive the water in an open channel. Since the only energy causing flow in an open channel is elevation energy, the gravitational force drives the water from higher to lower elevations.

The total available energy, E, of the still water of a lake or reservoir equals the elevation of the water surface with respect to any arbitrary reference level or datum plane below. This base elevation may be the mean sea level or any conveniently determinable elevation below the deepest point of the channel to be analyzed. In a flowing river or other kind of open channel the total available energy with respect to our reference base elevation below the bottom of the channel may be separated into three distinct components. These are the relative elevation Y of the bottom above the base line, the depth y of the water, and the kinetic energy head $v^2/2g$ plotted above the water surface.

In channels other than those of rectangular shape the depth of the water is variable in each cross section along the flow line. For this reason it is convenient for computational reasons to introduce the concept of average depth. Average depth is computed by dividing the cross-sectional area of the flow by the width of the water surface, B. Both of these values are to be measured in a plane that is perpendicular to the flow line, that is, to the direction of the velocity. Hence, the average depth is defined as

$$y_{ave} = \frac{A}{B} \qquad (8.1)$$

By such averaging irregular channels may be considered as rectangular in shape. The velocity in any channel is variable across the flow area because of the boundary layer effects along the channel boundaries. In practice the magnitude and direction of the velocity measured at various points in the cross-sectional area are rarely found to be uniform and parallel with the direction of the channel. For the purposes of hydraulic computations an average flow velocity is used; this is defined as the discharge of the flow divided by its cross-sectional area, that is,

$$v_{ave} = \frac{Q}{A} \qquad (8.2)$$

In practice this velocity was found to be about the average value of the velocities measured at 0.2 and 0.8 of the depth, y.

$$v_{ave} = \frac{v_{(at\,0.2y)} + v_{(at\,0.8y)}}{2} \qquad (8.3)$$

In many cases a good approximation of the average flow velocity is at 60% of the depth measured from the water surface.

EXAMPLE 8.1

A channel of nonuniform cross section as shown below delivers 100 cfs. The channel width is measured 30 ft at the water level (Figure E8.1). With an estimated average channel depth of 5 ft, estimate the average velocity.

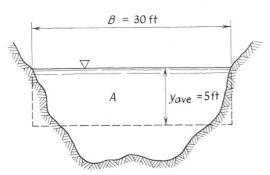

$B = 30$ ft

A $y_{ave} = 5$ ft

FIGURE E8.1

SOLUTION:

From Equation 8.1 the average depth

$$y_{ave} = \frac{A}{B}$$

Then
$$A = B \cdot y_{ave}$$
$$= 30 \times 5 = 150 \, ft^2$$

Therefore the average velocity can be calculated from Equation 8.2:

$$v = \frac{Q}{A}$$
$$= \frac{100}{150} = 0.67 \, fps$$

EXAMPLE 8.2

A trapezoidal channel with a bottom width of 4 m and side slopes of 4 : 1 (four horizontal to one vertical) carries water at a depth of 2 m (Figure E8.2). Compute the average depth.

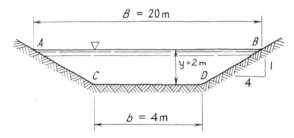

FIGURE E8.2

SOLUTION:

The cross-sectional area of the channel is
$$A = by + zy^2 = 4 \times 2 + 4 \times 2^2 = 24 \, m^2$$

The width of water surface, $B = 20$ m

The wetted perimeter, $P = 2\overline{AC} + \overline{CD}$

$$= 2\sqrt{2^2 + 8^2} + 4$$
$$= 20.49 \, m$$
$$z = 4$$
$$b = 4 \, m$$
$$B = 20 \, m$$

Then from Equation 8.1 the average depth is defined as

$$y_{ave} = \frac{A}{B}$$
$$= \frac{24}{20} = 1.2 \, m$$

EXAMPLE 8.3

If the discharge in the channel of Example 8.2 is 50 m³/s, calculate the average velocity.

SOLUTION:

The average velocity is defined as Equation 8.2.

$$v = \frac{Q}{A}$$

where

$$Q = 50 \, \frac{m^3}{s}$$

$$A = 24 \, m^2$$

therefore

$$v = \frac{50}{24}$$

$$= 2.08 \, \frac{m}{s}$$

In order to gain an insight into the complex relationships between energy, discharge, and depth in an open channel let us consider the *E total available energy* to be constant and measured with reference to the bottom of the channel at a certain cross section of rectangular shape and *B* width. In that case the total available hydraulic energy, often called "specific energy," may be written as

$$E = \frac{v^2}{2g} + y = \frac{Q^2}{2gB^2y^2} + y \qquad (8.4)$$

This equation may be plotted for a channel of unit width as shown in Figure 8.1, such that the discharge is on the horizontal coordinate toward the right, and the kinetic energy together with the depth is on the vertical coordinate. As for each depth the velocity is known, its value is also shown plotted on a horizontal coordinate toward the left. Proceeding from the top of the vertical coordinate downward we can draw a number of conclusions: In still water, when the velocity is zero, the total available energy equals y. The discharge is consequently also zero.

With a small amount of velocity causing

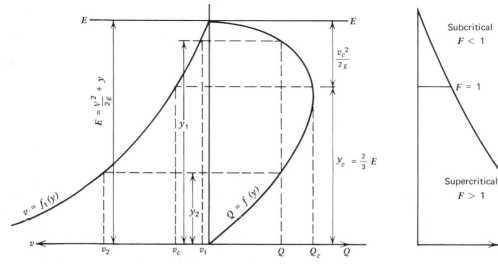

FIGURE 8.1 Velocity, depth, and discharge variation in open channel when the energy is constant.

flow, the depth of the water will be reduced by an amount corresponding to the kinetic energy head $v^2/2g$. For each definite amount of velocity there is a corresponding definite value of depth. As long as the total available energy remains the same, there is only one value of y for each value of v.

EXAMPLE 8.4

Calculate the total available energy for the problem described in Example 8.1.

SOLUTION:

By Equation 8.4 the total available energy is the sum of the depth and the kinetic energy term. Hence

$$E = \frac{Q^2}{2gy^2} + y = \frac{100^2}{2(32.2)5^2} + 5 = 11.21 \text{ ft}$$

Note that the total available energy is the hydraulic energy of the flow with respect to the bottom of the channel. In this problem the bottom is assumed to be at a uniform 5 ft below the water level.

CRITICAL FLOW CONDITION

With the increase of the velocity the discharge of the flow increases until it reaches a maximum value. This maximum discharge is called *critical discharge*, Q_c. The depth at which critical discharge occurs is called critical depth, the corresponding velocity is called critical velocity. By finding this extreme value using mathematical manipulations, we can prove that in rectangular channels Q_c occurs at the depth of

$$y_c = \tfrac{2}{3}E_{minimum} \qquad (8.5a)$$

This equation, of course, is only valid for rectangular cross sections. Substituting Equation 8.4 into 8.5a we could express the critical discharge as

$$\frac{Q_c}{B} = q_c = \sqrt{gy_c^3} \qquad (8.6a)$$

in the case of our rectangular channel of width B. Also, since at critical flow the kinetic energy head is one-half of the depth we find that

$$v_c = \sqrt{g \cdot y_c} \qquad (8.7a)$$

which, incidentally, is the equation giving the velocity of travel of small surface waves in shallow waters and is called *wave celerity*. If a small stone is dropped into the water, it will cause small ringlike waves around the point of impact. As long as the velocity of the water is less than critical, these rings will travel upstream. At critical velocity the ring waves will not be able to travel upstream since their velocity equals that of the stream itself. At velocities higher than critical the ringlets will not be able to form; they will be washed off, forming a v-shaped shockwave on the water surface.

In hydraulic computations the need often arises for the calculation of the critical depth y_c for a flow of given discharge Q at a known noncritical depth y. The question of this type is usually posed such: What would be the critical depth of a given flow if, while maintaining the same discharge, it would be made to flow at critical velocity? The solution of such problems is done by first calculating the average velocity of the flow at the given discharge and depth. The cross-sectional area at this depth has to be known to solve

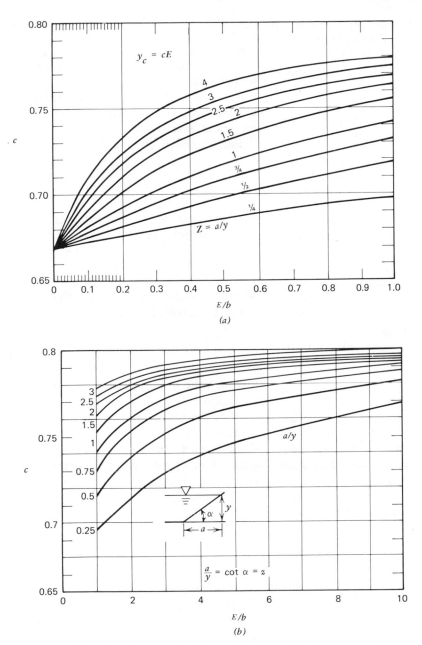

FIGURE 8.2 Graphs to calculate critical depths for trapezoidal channels.

the problem. Once the velocity is determined the kinetic energy $v^2/2g$ is calculated. This kinetic energy term when added to the depth of the flow will give the total available energy. Once this is known the critical depth is also available by virtue of Equation 8.5a if our cross section is rectangular.

For triangular cross sections the critical depth relates to the total available energy by the formula

$$y_c = \tfrac{4}{5}E \qquad (8.5b)$$

and, for parabolic cross sections,

$$y_c = \tfrac{3}{4}E \qquad (8.5c)$$

For partially full circular pipes the graph given in Appendix III provides an easy solution for the critical depth.

In trapezoidal cross sections the similar relationship takes the form

$$y_c = \frac{4zE - 3b + \sqrt{16z^2E^2 + 16zEb + 9b^2}}{10z}$$

(8.5d)

in which z is the side slope tabulated for various soils in Table 8-1, expressing the

Table 8-1 Allowable Side Slopes for Trapezoidal Channels in Various Soils

Type of Soil	z value
Loose clay or sandy loam	3
Loose silty sand	2
Firm clay	1.5
Earth with stone lining	1
Stiff earth with concrete lining	0.5–1
Muck and peat soils	0.25
Rock	0

horizontal measure on the slope for a unit measure of rise and b is the bottom width of the channel. Equation 8.5d is rather cumbersome to use. By introducing the ratio E/b, where E is the total available energy and b is the width of the channel bottom, Equation 8.5d could be solved as

$$y_c = cE \qquad (8.5d)$$

where c may be obtained from the graphs shown in Figure 8.2 as a function of E/b and the channel's side slope z.

The critical velocity in nonrectangular channels can be computed by the following equations:

For triangular channels

$$v_c = \sqrt{\frac{gy_c}{2}} \qquad (8.7b)$$

in which y_c is the result of Equation 8.5b.

For parabolic cross sections

$$v_c = \sqrt{\tfrac{2}{3}gy_c} \qquad (8.7c)$$

where y_c is the result of Equation 8.5c.

For trapezoidal channels

$$v_c = \sqrt{\frac{b + zy_c}{b + 2zy_c} \, gy_c} \qquad (8.7d)$$

in which the notations are the same as in Equation 8.5d.

Once the critical depth and the critical velocity is available for any of the above listed cross sections the critical discharge may be computed without further difficulty. Critical discharge formulas for various channel section geometries are included in Appendix III.

EXAMPLE 8.5

If the discharge in the trapezoidal channel described in Example 8.2 is 30 m^2/s what is the critical depth, the critical velocity, and the critical discharge of the flow?

SOLUTION:

The area of the channel was 24 m^2. Hence the average velocity is $v = Q/A = 30/24 = 1.25$ m/s.
The kinetic energy is

$$\frac{v^2}{2g} = \frac{1.25^2}{2(9.81)} = 0.0796 \text{ m}$$

The total available energy is then

$$E = \frac{v^2}{2g} + y = 0.0796 + 2.0 = 2.0796 \text{ m}$$

By Equation 8.5d and Figure 8.2 we need to express E/b, which is

$$\frac{E}{b} = \frac{2.0796}{4} = 0.52$$

which is in graph a of Figure 8.2 and for $z = 4$ gives $c = 0.768$. Hence

$$y_c = cE = 0.768(2.0796)$$
$$y_c = 1.59 \text{ m}$$

To obtain the critical velocity, we use Equation 8.7d:

$$v_c = \sqrt{\frac{b + zy_c}{b + 2zy_c} \, gy_c} = \sqrt{\frac{4 + 4(1.59)}{4 + 2(4)(1.59)} \, 9.81(1.59)} = 3.11$$

$$v_c = 3.11 \frac{\text{m}}{\text{s}}$$

To obtain the critical discharge we first compute the area corresponding to the critical depth

$$A_c = b \cdot y_c + 2\left(\frac{zy_c^2}{2}\right) = 4(1.59) + 2\left(\frac{4(1.59)^2}{2}\right)$$
$$6.36 + 10.11 = 16.47 \text{ m}^2$$

The critical discharge is then equal to

$$Q_c = A_c v_c = 16.47(3.11) = 51.23 \frac{\text{m}^3}{\text{s}}$$

EXAMPLE 8.6

A rectangular channel of 12 ft width has a critical depth of 4.5 ft. Compute the corresponding critical discharge.

SOLUTION:

The cross-sectional area

$$A_c = B \cdot y_c = 12(4.5) = 54 \text{ ft}^2$$

By Equation 8.7a,

$$v_c = \sqrt{gy_c} = \sqrt{32.2(4.5)} = 12.04 \text{ fps}$$

The discharge is

$$Q_c = A_c \cdot v_c = 54(12.04) = 650 \text{ cfs}$$

NONCRITICAL FLOW CONDITIONS

Under a given total available energy at one point, any discharge less than Q_c may flow at any two possible depths. As shown by the vertical line Q crossing the discharge curve in Figure 8.1, these two depths are predetermined by virtue of the energy equation, Equation 8.4. One of these depths is always greater, the other one is less than the critical depth. The two corresponding depths are called *conjugate depths*. If one is known, the other can be calculated by writing the sum of the depth and kinetic energy terms for both cases and equating these. For rectangular channels this yields

$$y_2 = \frac{y_1}{2}\left[-1 + \sqrt{1 + \frac{8q^2}{gy_1^3}}\right] \quad (8.8a)$$

where q equals Q/B. In case the velocity is less than critical the flow is called subcritical or tranquil. Most river flows as well as flows in manmade channels are tranquil flows. When the velocity is higher than critical, it is called a supercritical or rapid flow. Flows over spillways, in chutes, in steep mountain creeks, or in rain water rushing toward storm inlets on pavements the flow is supercritical.

The question whether a flow is subcritical, critical, or supercritical is important enough to warrant the introduction of a special dimensionless parameter, called *Froude number*. The Froude number is formed by taking the ratio of the kinetic energy head to the depth of the flow in such manner that the number is unity at critical flow. Since at critical flow

$$\frac{v_c^2}{2g} = \frac{y_c}{2}$$

The Froude number $\mathbf{F}$ may be defined as

$$\mathbf{F} = \frac{v^2}{gy} \quad (8.9a)$$

which is now a dimensionless term independent of the units used in the computations. In hydraulics literature as well as engineering reports the Froude number is often defined as the square root of Equation 8.9a,

$$\mathbf{F} = \frac{v}{\sqrt{gy}} = \frac{\text{velocity}}{\text{wave celerity}} \quad (8.9b)$$

To prevent confusion in case of noncritical flows, whenever reference is made to the Froude number it is important to clarify the form in which it is expressed. As shown in Figure 8.1, the value of the Froude number is less than one for subcritical flows, equals one at critical flow,

and is greater than one for supercritical flows.

In rectangular channels and some large rivers Equation 8.9a may be written in the form

$$\mathbf{F} = \frac{1}{B^2} \cdot \frac{Q^2}{gy^3} \quad (8.10)$$

where Q is the discharge, B is the width of the channel, and y is the depth of the flow. The nomograph shown in Figure 8.3 will facilitate the computation of Equation 8.10 when English units are used.

Small natural watercourses and some artificial channels are better approximated by a parabolic cross section rather than a rectangular one. For parabolic channels the Froude number of the form of Equation 8.9a can be computed by

$$\mathbf{F} = 0.296 \frac{Q^2}{gy^4} \quad (8.11)$$

where y is the maximum depth. Equation 8.11 is plotted in nomograph form in Figure 8.4 using English units. By observing Equations 8.10 and 8.11 we may note that they could be generalized in some form. For critical flow $\mathbf{F} = 1$ and $Q = Q_c$ and the equations could be put in a form of

$$\frac{Q_c^2}{g} = f(y^M) \quad (8.12)$$

in which M is an exponent that is constant for a particular channel shape. M is called *critical flow exponent*. For common shapes of artificial channels M may be found on the graph shown in Figure 8.5. The variable z for trapezoidal channels built in earth depend on slope stability. Design values of z for different soils were given in Table 8-1.

Under certain conditions supercritical flow may suddenly change into subcritical flow. This sudden change is called a *hydraulic jump*. It occurs after a stream of water rushes down a spillway, particularly if the downstream conditions are such that they impede further supercritical flow, as by the presence of energy absorbing devices or by a high water level pool.

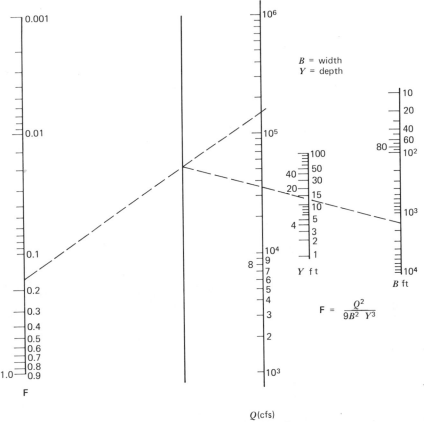

$$Q(\text{cfs})$$

FIGURE 8.3 Froude number of flow in rectangular channel. (Equation 8.9a.)

The hydraulic jump means that the water level suddenly changes from the supercritical conjugate depth y_1 to the subcritical conjugate depth y_2. Equation 8.8a shows the correspondence between these two depths, but for practical use it is convenient to convert this equation by introducing the Froude number for the supercritical flow, F_1 as a variable, in the form of Equation 8.9b. This mathematical conversion results in a convenient dimensionless formula,

$$\frac{y_2}{y_1} = \frac{1}{2}(\sqrt{1 + 8F_1^2} - 1) \qquad (8.8b)$$

Depending on the magnitude of the Froude number at the beginning of the jump, we may note that the strength of the jump increases with F_1. For F_1 less than 1.7, experiments indicate that the jump appears as a series of undulating waves along the downstream surface. Between 1.7 and 2.5 a series of rollers appear on the surface, but the presence of the jump does not significantly disturb the downstream water level and it is considered weak and unstable in its position. For Froude numbers of 2.5 to 4.5 the jump grows in strength, causing forceful random waves to travel downstream. For this range the elevation difference between the two sides of the jump is the greatest. For 4.5 to 9.0 the position of the jump stabilizes in the channel and is less influenced by the tailwater depth. For F_1 larger than 9.0 the jump again causes downstream waves, but their effect is less destructive as the energy dissipation in such jumps is high, because a hydraulic jump is associated with high turbulence, vortices, and swirls. As a result there is considerable loss of total available energy in hydraulic jumps.

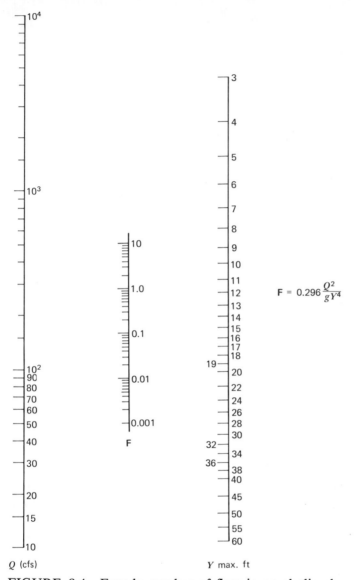

FIGURE 8.4   Froude number of flow in parabolic channel. (Equation 8.9a.)

The energy loss through a hydraulic jump may be computed by the formula

$$E_{loss} = \frac{[y_2 - y_1]^3}{4 y_1 y_2} \qquad (8.13)$$

The energy absorbing character of the hydraulic jump will be further discussed in the following chapter.

As the discharge and the two conjugate depths are inherently related by Equation 8.8, the relationship can be utilized to determine the discharge by measuring the depths before and after the hydraulic jump in a rectangular channel and substituting into

$$q = \frac{Q}{B} = \left[\frac{y_1 + y_2}{2} (g y_1 y_2) \right]^{1/2} \qquad (8.14)$$

where q is the discharge for a unit width of channel. There is no corresponding sudden change of water level when the flow changes from subcritical to supercritical. Hence there is no great energy loss associated with such changes. Hydraulic jump is only possible when the velocity is reduced from supercritical to subcritical.

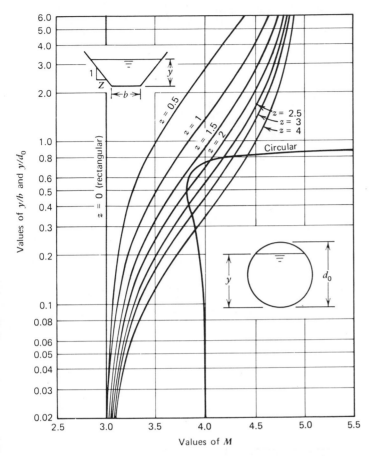

FIGURE 8.5 Critical flow exponents for rectangular, trapezoidal, and circular channels. (From *Open Channel Hydraulics* by Ven Te Chow. Copyright 1958, McGraw-Hill Book Co., Inc. Used with permission of McGraw-Hill Book Company.)

EXAMPLE 8.7

On a 12 m wide rectangular chute water flows in supercritical condition at a rate of 150 m³/s. At the end of the chute, on a horizontal concrete apron the pool level of the downstream water is at 3 m above the apron. This downstream water will cause the formation of a hydraulic jump. Analyze the jump.

SOLUTION:

The discharge per unit width is

$$q = \frac{Q}{B} = \frac{150}{12} = 12.5 \frac{\text{m}^3}{\text{s}}$$

The Froude number at the front of the jump, written in the form of Equation 8.9b, is

$$\mathbf{F}_1 = \frac{v}{\sqrt{gy_1}} = \frac{q/y_1}{\sqrt{gy_1}} = \frac{q}{\sqrt{g}\ y_1^{3/2}}$$

Using Equation 8.8b to express the conjugate depth y_1 for the known $y_2 = 3$, we have

$$\frac{y_2}{y_1} = \frac{1}{2}(\sqrt{1 + 8F_1{}^2} - 1)$$

$$= \frac{1}{2}\left(\sqrt{1 + 8\frac{q^2}{gy_1{}^3}} - 1\right)$$

From this, we get

$$y_2 y_1{}^2 + y_1 y_2{}^2 - \frac{2q^2}{g} = 0$$

and, substituting all known values

$$3y_1{}^2 + 9y_1 - \frac{2(12.5)^2}{9.81} = 0$$

or

$$y_1{}^2 + 3y_1 - 10.62 = 0$$

Solving this quardratic equation we find that

$$y_1 = 2.09 \text{ m}$$

Substituting into the expression for the Froude number we get

$$F = \frac{q}{\sqrt{g}\ y_1{}^{3/2}} = \frac{12.5}{\sqrt{9.81}\ (2.09)^{3/2}} = 1.32$$

which is less than 1.7. As described on page 151 the hydraulic jump will take the form of a series of undulating waves along the downstream water surface. The location of the jump is rather ill defined and shore protection is required due to the wave action.

The energy loss in the jump may be computed from Equation 8.13 as

$$E_{\text{loss}} = \frac{(y_2 - y_1)^3}{4y_1y_2} = \frac{(3 - 2.09)^3}{4(3)\ 2.09} = 0.03 \text{ m}$$

The total available energy downstream, by Equation 8.4 is

$$E_2 = y + \frac{q^2}{2gy_2{}^2}$$

$$E_2 = 3 + \frac{12.5^2}{2(9.81)3^2} = 3.884 \text{ m}$$

EXAMPLE 8.8

The rectangular channel 10 ft wide delivers 320 cfs of water at 1.8 ft depth before entering a hydraulic jump (Figure E8.8). Compute the downstream water level and the critical depth.

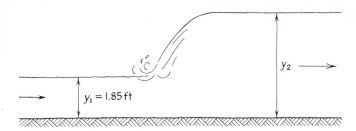

FIGURE E8.8

SOLUTION:

$$q = \frac{\text{Discharge}}{\text{unit width}} = \frac{320}{10}\frac{\text{cfs}}{\text{ft}}$$

$$q = 32\frac{\text{cfs}}{\text{ft}}$$

$$y_1 = 1.8\,\text{ft}$$

From Equation 8.8a

$$y_2 = \frac{y_1}{2}\left[\sqrt{1 + \frac{8q^2}{gy_1^3}} - 1\right]$$

$$= \frac{1.8}{2}\left[\sqrt{1 + \frac{8(32)^2}{32.2 \times (1.8)^3}} - 1\right]$$

$$y_2 = \text{downstream water level} = 5.11\,\text{ft}$$

To find the critical depth; from Equation 8.6a,

$$y_c = \left(\frac{q^2}{g}\right)^{1/3}$$

$$= \left[\frac{(32)^2}{32.2}\right]^{1/3}$$

$$y_c = 3.17\,\text{ft}$$

EXAMPLE 8.9

The downstream depth after a hydraulic jump in a 12 ft wide rectangular channel is 5 ft. Compute a rating curve showing the discharges and the corresponding conjugate upstream depths.

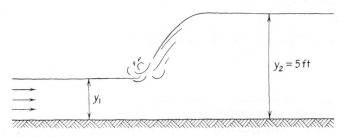

FIGURE E8.9

SOLUTION:

From Equation 8.14,

$$q = \left[\frac{y_1 + y_2}{2}(gy_1y_2)\right]^{1/2}$$

$$= \frac{\text{discharge}}{\text{unit width}} = \frac{Q}{B}$$

If we substitute

$$Q = 12\left[\frac{y_1 + 5}{2}(32.2 \times 5y_1)\right]^{1/2}$$

$$= 107.666[y_1(y_1 + 5)]^{1/2}$$

y_1 ft	1.0	1.5	2.0	2.5	3.0	3.5	4.0	4.5	5.0
Q cfs	263.73	336.17	402.85	466.21	527.45	587.25	645.0	704.0	761.31

EXAMPLE 8.10

Determine the critical flow exponent for a trapezoidal channel of 4:1 side slopes and 16 ft bottom width when the depth of the flow is 3 ft (Figure E8.10).

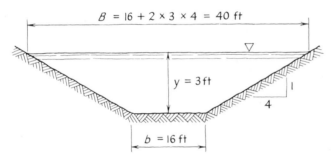

FIGURE E8.10

SOLUTION:

From the relation (8.12)

$$\frac{Q_c^2}{g} = f(y^M)$$

$$M = \text{critical flow exponent}$$

From Figure 8.5 the relation of z and y/b determines M: We have $y/b = 3/16 = 0.1875$, $z = 4$. Entering Figure 8.5 we obtain $M = 3.625$.

EXAMPLE 8.11

Compute the differences between corresponding critical flow exponents if the depth in Example 8.10 changes from 3 ft to 6 ft and 9 ft.

SOLUTION:

y	6	9
y/b	0.375	0.563
z	4	4
M	4.05	4.25

M values were determined from Figure 8.5 corresponding to y/b and z. Note how small the change in M is as compared to change in depth.

NATURAL CONDITIONS OF STEADY FLOW

Water in an open channel flows from the high point to the low point along the watercourse. The flowing water must overcome boundary induced resistances because water is a viscous fluid. Hence the flowing water gradually loses its energy along its path. If the energy loss along the channel is less than the energy gained by the decreasing elevation, the energy content of the discharge increases; therefore the flow is accelerating. Conversely, if the frictional losses along the channel exceed the energy gained by the downward flow, the flow is decelerating.

In case of a constant discharge, acceleration along the channel results in decreasing depth and deceleration results in increasing depth. In either case the water surface will not remain straight and parallel with the bottom of the channel but will be slightly curved. Examples of various water surface shapes for different flow conditions are depicted in Figures 8.6 and 8.7.

In the situations shown in these two illustrations the water flows from an upper to a lower reservoir through a channel that has an arbitrary but constant configuration along the flow path. Channels of this type are called *prismatic channels*. In both cases the channel bottom is of constant slope, and the water surface elevations are also the same regardless of the flow in the channel. In practice the case may represent a channel connecting a lake to a larger river.

In Figure 8.6 the depth of the water in the upper reservoir is always at the same elevation while the depth in the lower reservoir is varied for the sake of illustrating the principle in question. As long as

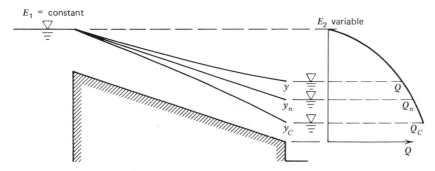

FIGURE 8.6   Discharges and water profiles in a finite channel with different downstream water levels.

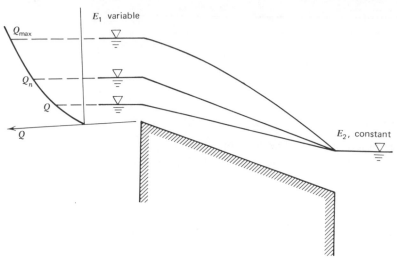

FIGURE 8.7   Discharges and water profiles in a finite channel with different upstream water levels.

the water level in the lower reservoir is the same as that in the upper reservoir no flow will take place in the channel. If the water level is reduced in the lower reservoir, the water will begin to flow. The energy gain causing the flow depends strictly on the difference between the water surface elevations in the reservoirs. The slope of the energy grade line, S_e, in the channel may be approximated by

$$S_e = \frac{H_1 - H_2}{L} = \frac{h_L}{L} \qquad (8.15)$$

where L is the length of the channel, H_1 is the elevation of the upper reservoir, H_2 is the elevation of the lower reservoir and h_L is what we call head loss or energy loss along L. Looking at the drawing we can recognize that as long as S_e is not parallel to S, the bottom slope of the channel, the available energy in the flow at any particular point along the channel will be different. Since the discharge along the channel is constant, the kinetic energy and the depth of the flow will vary from point to point, resulting in a curved water surface connecting the upper reservoir level to the lower reservoir level.

As the water level in the lower reservoir is lowered, more and more energy is made

available to cause flow. This increases the flowing discharge. With the upper water level remaining constant the discharge in the channel reaches a maximum value when the depth of the flow at the lower end of the channel reaches critical depth. For this case, when the flow is controlled by a critical condition at the exit the flow in the channel will depend on the position of the water level at the entrance of the channel. There are two possible cases. One is when the depth in the upstream reservoir results in an entrance depth in the channel that is less than the critical depth based on the exit condition. In this case the flow throughout the channel will be supercritical, with the depth of the water gradually rising because of deceleration caused by viscous shear losses. The other possible case arises when the upper reservoir level is such that the entrance depth is larger than the critical depth at the downstream end. In this case the flow throughout the channel will be subcritical, with the depth of the water slowly subsiding along the flow, characterizing an accelerating condition.

For a given discharge there is a particular case in which the water flows down the channel at critical flow conditions throughout. The slope at which this occurs is called *critical slope*. As critical flow is very unstable, flow on a critical slope is wavy and irregular, causing erosion, sediment deposition, etc. Critical or near critical slopes should be avoided in design as much as possible. Channel bottom slopes that are less than critical are referred to as *mild slopes*. Channels with slopes larger than critical are called *steep slopes*. All of these, of course, are dependent on the discharge flowing. A slope may be steep for one discharge, critical for another, and mild for an even larger discharge.

Figure 8.7 shows physical conditions identical to those depicted in Figure 8.6 with the only difference that now the water level in the lower reservoir is kept constant and the level in the upper reservoir is varied. In this case no flow will occur in the channel regardless of the lower reservoir level until the level in the upper reservoir exceeds the channel bottom's highest point. From there the flow will increase rapidly with the growth of the upper reservoir depth until the channel reaches its maximum capacity at which the flow into the lower reservoir is critical.

From here the problem gets more complex and as such is beyond the scope of this book. One aspect complicating the problem is that the flow can reach critical state at any other point of the channel depending on whether or not the bottom slope is mild, critical, or steep. The other complicating aspect is that in nature neither end of the channel is endowed with a constantly fixed water level. This results in an increase of the variables controlling the flow. Therefore the *channel delivery graph* is composed of a family of curves, each depicting a particular set of circumstances controlling the flow. The third complicating aspect is that in nature the bottom slope, or even the channel cross section, is varying along the path of the flow. The last complicating aspect arises because in practice the discharge in the channel rarely remains constant in time. Computations related to unsteady discharges are in the realm of flood routing, a science of considerable complexity.

THE CONCEPT OF NORMAL FLOW

In regard to the computational difficulties associated with the analysis of flow in open channels, engineers seeking a simple method for discharge calculations have developed formulas for the case when the energy grade line is assumed to be parallel with the bottom slope of the channel. In this simplified case the energy gained by the water at any point equals exactly the energy lost through friction. Since there is no acceleration or deceleration along the channel, the depth as well as the kinetic energy term $(v^2/2g)$ remains constant. Thus it is clear that this zero acceleration condition rarely occurs in practice, but the savings in computation difficulties as well as the uncertainty associated with the determination of the true expected discharge undoubtedly makes the approach worthwhile. The flow with no acceleration or deceleration is called *normal flow*.

Normal flow in open channels is computed by the Chézy formula which states that

$$v = \frac{Q}{A} = C\sqrt{RS} \qquad (8.16)$$

where

C = Chézy's resistance coefficient

v = average velocity

Q = discharge of the flow

A = cross-sectional area of the channel

S = slope of the channel $(S = S_e)$

The depth corresponding to such flow is called normal depth y_n. R in Equation 8.16 is called hydraulic radius, which is

$$R = \frac{A}{P} \qquad (8.17)$$

where P is the wetted perimeter of the

channel, that is, the length of the line along which the water is in contact with the channel bottom in the cross-sectional area A of the flow. In reality R is not a radius at all. It is rather a geometric parameter indicating the efficiency of the cross section; A is a positive and P a negative contributor to the movement of the water, since the more bottom surface is present to create frictional resistance the more the flow is retarded. Conversely, the larger the flow area as compared to P, the easier will the water move. In channel design an attempt is made to select the best possible hydraulic radius. Selection of the optimum hydraulic radius for trapezoidal or other channels is a rather cumbersome procedure. A useful graphical aid for general channel design, including the determination of the channel shape, is shown in Figure 8.8. This graph, developed by T. Lenkei and used extensively in Hungary, allows the solution of the Chézy formula completely for trapezoidal channels with channel side slopes of $z = 1, 1.5,$ and 2. The three lower graphs in Figure 8.8 show curves of constant hydraulic radius in the coordinates of the depth and bottom width of these three trapezoidal shapes. The radial lines emanating from the origin of these coordinates refer to the optimum hydraulic radius, the 3% deviation from this optimum, as well as hydraulic radius values for channels of given depth to bottom width ratios. As soon as the value of the hydraulic radius was selected with the aid of these graphs, it may be located on the vertical coordinate of the upper two graphs. The left one of these is labeled channel shape graph; the right one is the channel slope graph. After selecting the appropriate lines on these two graphs, the horizontal distance between the two lines gives the discharge carried by the channel at any vertical elevation corresponding to a hydraulic radius. The magnitude of this discharge is determined by the discharge scale drawn in the middle of Figure 8.8. The horizontal coordinate of the channel slope graph is the average velocity of the

flow. The families of the lines shown in this graph are constant channel slope values for earth channels and channels with concrete and rock linings. The horizontal coordinate of the channel shape graph allows the direct determination of the corresponding cross-sectional area for semicircular, various trapezoidal, and rectangular channels of a given hydraulic radius.

In the general practice the determination of the design depth of the channel is sometimes simplified by neglecting optimum conditions for the hydraulic radius and making the selection of the depth y a function of the cross-sectional area A alone. Such formulas may be written generally as

$$y = \sqrt{\frac{A}{e}} \qquad (8.18)$$

where the constant e is taken to be 4 by the U.S. Bureau of Reclamation and 3 in irrigation channel design practice in India.

Usually for most channels the bottom width b is predetermined and y is calculated from the equations. In some cases when depth considerations are important, such as when the water level is to be kept near the ground water level as in the case of drainage and swamp reclamation design, the depth is predetermined and the base width b of the channel is computed. In either case the graph in Figure 8.8 is a useful design aid for trapezoidal channels. In wide rivers where the surface width B is ten or more times larger than the depth d, the hydraulic radius is assumed to equal the depth of the flow, that is $R = d$.

C in Equation 8.16 is Chézy's channel resistance coefficient, an experimentally determined factor with the dimension shown by Equation 10.37. Based on a great deal of measurements in the field as well as in laboratory channels since the beginning of the eighteenth century the value of C was determined in metric units to be

$$C = \frac{1}{n} R^{1/6} \qquad (8.19)$$

in which n is the so-called Kutter's coeffi-

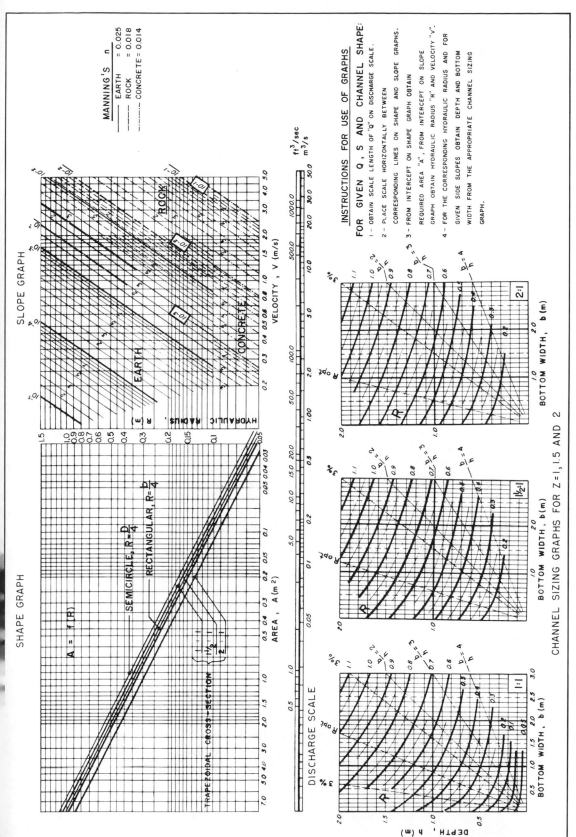

FIGURE 8.8 Lenkei's channel design diagram (Courtesy of Tibor Lenkei, Budapest, Hungary).

cient, a resistance factor referring to the condition of the channel. Table 8-2 shows values of n for various channel conditions. Equation 8.19 is known in the literature as Manning's formula. In English in contrast to metric units, Equation 8.19 is written as

$$C = \frac{1.49}{n} R^{1/6} \qquad (8.20)$$

with C now in ft/sec and R in ft. The number 1.49 is the conversion factor,

enabling the user to use Kutter's metric n values without unit conversion. Combining Equations 8.16 and 8.19 one obtains the so-called Chézy-Manning formula in metric units

$$Q = \frac{A}{n} R^{2/3} S^{1/2} \qquad (8.21)$$

giving Q in m³/s. R and A are to be substituted in m and m², respectively. The equation serves as a basis for Figure 8.8.

Table 8-2   Roughness Coefficients for Open Channels

Description of Channel	n	$\dfrac{1.49}{n}$
Exceptionally smooth, straight surfaces: enameled or glazed coating; glass; lucite; brass	0.009	165.55
Very well planed and fitted lumber boards; smooth metal; pure cement plaster; smooth tar or paint coating	0.010	149.00
Planed lumber; smoothed mortar ($\frac{1}{3}$ sand) without projections, in straight alignment	0.011	135.45
Carefully fitted but unplaned boards, steel troweled concrete, in straight alignment	0.012	124.16
Reasonably straight, clean, smooth surfaces without projections; good boards; carefully built brick wall; wood troweled concrete; smooth, dressed ashlar	0.013	114.62
Good wood, metal, or concrete surfaces with some curvature, very small projections, slight moss or algae growth or gravel deposition. Shot concrete surfaced with troweled mortar	0.014	106.43
Rough brick; medium quality cut stone surface; wood with algae or moss growth; rough concrete; riveted steel	0.015	99.33
Very smooth and straight earth channels, free from growth; stone rubble set in cement; shot, untroweled concrete; deteriorated brick wall; exceptionally well excavated and surfaced channel cut in natural rock	0.017	87.65
Well-built earth channels covered with thick, uniform silt deposits; metal flumes with excessive curvature, large projections, accumulated debris	0.018	82.77
Smooth, well-packed earth; rough stone walls; channels excavated in solid, soft rock; little curving channels in solid loess, gravel or clay, with silt deposits, free from growth, in average condition; deteriorating uneven metal flume with curvatures and debris; very large canals in good condition	0.020	74.50
Small, manmade earth channels in well-kept condition; straight natural streams with rather clean, uniform bottom without pools and flow barriers, cavings and scours of the banks	0.025	59.60
Ditches; below average manmade channels with scattered cobbles in bed	0.028	53.21

Table 8-2 *Cont'd*

Description of Channel	n	$\dfrac{1.49}{n}$
Well-maintained large floodway; unkept artificial channels with scours, slides, considerable aquatic growth; natural stream with good alignment and fairly constant cross section	0.030	49.66
Permanent alluvial rivers with moderate changes in cross section, average stage; slightly curving intermittent streams in very good condition	0.033	45.15
Small, deteriorated artificial channels, half choked with aquatic growth, winding river with clean bed, but with pools and shallows	0.035	42.57
Irregularly curving permanent alluvial stream with smooth bed; straight natural channels with uneven bottom, sand bars, dunes, few rocks and underwater ditches; lower section of mountainous streams with well-developed channel with sediment deposits; intermittent streams in good condition; rather deteriorated artificial channels, with moss and reeds, rocks, scours, and slides	0.040	37.25
Artificial earth channels partially obstructed with debris, roots, and weeds; irregularly meandering rivers with partly grown-in or rocky bed; developed flood plains with high grass and bushes	0.067	22.24
Mountain ravines; fully ingrown small artificial channels; flat flood plains crossed by deep ditches (slow flow)	0.080	18.62
Mountain creeks with waterfalls and steep ravines; very irregular flood plains; weedy and sluggish natural channels obstructed with trees	0.10	14.9
Very rough mountain creeks, swampy, heavily vegetated rivers with logs and driftwood on the bottom; flood plain forest with pools	0.133	11.2
Mudflows; very dense flood plain forests; watershed slopes	0.22	6.77

Combining Equation 8.16 with Equation 8.20 gives the Chézy-Manning formula in English units:

$$Q = \frac{1.49}{n} AR^{2/3} S^{1/2} \qquad (8.22)$$

with Q in cfs and A and R in ft^2 and ft, respectively. Figure 8.9 shows a popular nomograph by which Equation 8.22 may be solved.

In the case of normal flow in partially full circular channels, the computations are facilitated by the graph given in Appendix III. In both Equations 8.21 and 8.22 the value of S is the elevation difference between the two ends of the channel divided with the channel length. In cases of design, S may be fixed by a predetermined alignment of the channel, such as for roadside ditches. Even in these cases reduction of the slope is possible occasionally by using drop structures along the channel at certain intervals. In other cases when only the two ends of the channel are fixed in space, the slope can be decreased at will by following land contours with a predetermined slope S in which case the length of the channel will be determined from

$$L = \frac{H_1 - H_2}{S} \qquad (8.23)$$

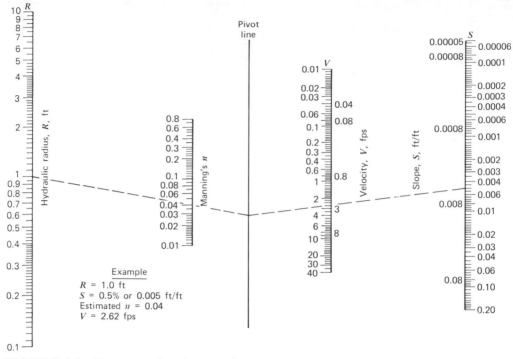

FIGURE 8.9   Nomograph solution of the Chézy-Manning equation.

where H_1 is the upper available energy line elevation, H_2 is the elevation of the water level at the recipient water body at the lower end of the channel, and S is the predetermined channel slope. L of course enters into the design considerations as an economic factor, giving rise to the total required amount of land acquisition and excavation along with the required cross-sectional area A and the so-called "free board," the depth of the water level below the top of the channel. The value of the *free board*, u, may be determined by

$$u = \sqrt{c \cdot y} \qquad (8.24)$$

in which the freeboard is obtained in feet for a given depth of water y in feet. The U.S. Bureau of Reclamation recommends that the factor c selected to be 1.5 for a small canal of 20 cfs capacity and 2.5 for canal capacities of 3000 cfs or more.

For a general cross section the Chézy-Manning equation may be written as

$$\frac{Q}{\sqrt{S}} = K \qquad (8.25)$$

where K is called the *conveyance*. By Equation 8.21

$$K = \frac{1}{n} AR^{2/3} \qquad (8.26)$$

in which both A and R depend on the geometry. For any given channel geometry, Equation 8.25 can be expressed by the channel depth y such that

$$K^2 = my^N = f(y^N) \qquad (8.27)$$

where m is a coefficient depending on roughness and geometry and N is called the *normal flow exponent*. Equation 8.27 is comparable to Equation 8.12, which expresses a similar concept for critical flow. Both of these hydraulic exponents are important parameters in design when normal flow cannot be assumed to develop. Values of N are shown in Figure 8.10 for common trapezoidal and circular channels.

If the discharge in a channel as well as its bottom slope and roughness are known, both the normal depth and the

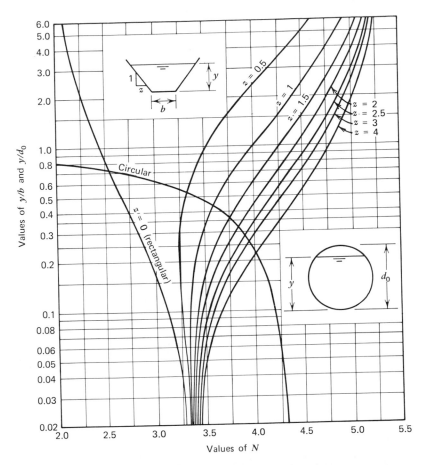

FIGURE 8.10 Normal flow exponents for rectangular, trapezoidal and circular channels. (From *Open Channel Hydraulics* by Ven Te Chow. Copyright 1958, McGraw-Hill Book Co., Inc. Used with permission of McGraw-Hill Book Company.)

critical depth can be calculated from formulas presented above. If the critical depth is shown to be smaller than the normal depth the normal flow is subcritical. Conversely, if the critical depth is larger than the normal depth, the normal flow is supercritical. The slope at which the critical depth equals the normal depth is the critical slope, discussed earlier in this chapter. Most natural watercourses as well as manmade channels conduct the flow under subcritical conditions.

EXAMPLE 8.12

The slope of the channel described in Example 8.2 is $S = 0.002$ (Figure E8.12). Determine the roughness coefficient n of the channel assuming that normal flow is encountered.

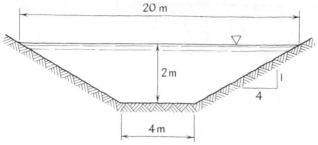

FIGURE E8.12

SOLUTION:

$$\text{Area} = 24 \text{ m}^2$$

$$Q = 50 \frac{\text{m}^3}{\text{s}}$$

$$V = 2.08 \frac{\text{m}}{\text{s}}$$

$$R = \frac{24}{20.49} = 1.17 \text{ m}$$

$$S = 0.002$$

From Equation 8.16

$$v = \frac{Q}{A} = C\sqrt{RS}$$

Therefore, Chézy's coefficient

$$C = \frac{2.08}{\sqrt{1.17(0.002)}} = 43$$

From Equation 8.19

$$C = \frac{1}{n} R^{1/6}$$

$$n = \frac{R^{1/6}}{C}$$

$$= \frac{(1.17)^{1/6}}{43} = 0.024$$

A roughness coefficient of 0.024 in Table 8-2 corresponds to small manmade earth channels in well-kept condition.

EXAMPLE 8.13

For a triangular channel of $z = 2$, side slopes carrying a discharge of 100 cfs at a depth of 5 ft determine the channel slope S under the assumption that normal flow exists and that n equals 0.022.

SOLUTION:

The surface width B is

$$B = 2z \cdot y = 2(2)5 = 20 \text{ ft}$$

The cross-sectional area A, hence, is

$$A = \frac{1}{2} B \cdot y = \frac{1}{2}(20)5 = 50 \text{ ft}^2$$

The velocity

$$v = \frac{Q}{A} = 2 \text{ fps}$$

The hydraulic radius from Equation 8.17 is

$$R = \frac{A}{P}$$

$$P = \sqrt{\left(y^2 + \frac{B}{2}\right)^2} = 2\sqrt{25 + 100} = 22.36 \text{ ft}$$

$$R = \frac{50}{22.36} = 2.23 \text{ ft}$$

From Equation 8.20

$$C = \frac{1.49}{n} R^{1/6} = \frac{1.49}{0.022} 2.23^{0.166} = 77.4 \ \sqrt{\text{ft/sec}}$$

Entering Equation 8.16, we get

$$v = C\sqrt{R \cdot S}$$

$$2 = 77.4\sqrt{2.23S}$$

From this equation

$$S = \left(\frac{2}{77.4}\right)^2 \frac{1}{2.23} = 3 \times 10^{-4} \frac{\text{ft}}{\text{ft}}$$

or 3 ft of drop over 10,000 ft.

EXAMPLE 8.14

A trapezoidal earth channel of $1:1$ side slopes is to be built on a slope of $S = 0.001$ to carry $1.08 \text{ m}^3/\text{s}$ water. Design the channel cross section such that the hydraulic radius is optimal.

SOLUTION:

Using Figure 8.8 first we mark off the length of the $1.08 \text{ m}^3/\text{s}$ discharge on the edge of a sheet of paper. Next, keeping the line horizontal we place the paper's edge on the upper graphs, moving it upward along the corresponding slope ($S = 0.001$) and shape ($1:1$) lines. Where the distance between the lines equals the discharge length we note the magnitude of the hydraulic radius R. Carrying out the suggested procedure, we find that

$$R = 0.45 \text{ m}$$

The corresponding velocity in the channel is

$$v = 0.75 \frac{\text{m}}{\text{s}}$$

obtained from the slope graph. The cross-sectional area is

$$A = 1.5 \text{ m}^2$$

from the shape graph.

Entering the left bottom graph along the $R = 0.45$ curve, we find the intercept with the radial line indicating optimum conditions. In our case for

$$R = R_{opt} = 0.45 \text{ m}$$

we have

$$b = 0.8 \text{ m bottom width}$$

and

$$y = 0.9 \text{ m water depth}$$

EXAMPLE 8.15

For the open channel flow problem treated in Example 8.12, determine the normal flow exponent N.

SOLUTION:

Use Figure 8.10 with

$$b = 4 \text{ m}$$
$$y = 2 \text{ m}$$
$$\frac{y}{b} = \frac{2}{4} = 0.5$$
$$z = 4$$

Entering the graph at the vertical coordinate at 0.5 and turning down at the $z = 4$ curve, we find

$$N = 4.5$$

EXAMPLE 8.16

A trapezoidal channel of bottom width 20 ft side slope $z = 3$ carries a flow of 400 cfs at normal conditions with a depth of 3.3 ft from a point with elevation of 786 ft above sea level to a recipient water body of 680 ft water level (Figure E8.16). Determine the required channel length, if $n = 0.025$.

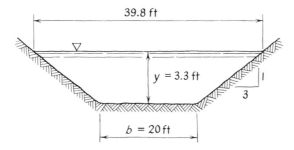

FIGURE E8.16

SOLUTION:

$$Q = 400 \text{ cfs}$$

$$z = 3$$

$$y = 3.3 \text{ ft}$$

$$b = 20 \text{ ft}$$

$$n = 0.025$$

$$B = 20 + 2 \times 3 \times 3.3 = 39.8 \text{ ft}$$

Then

$$A = \frac{1}{2}(39.8 + 20)3.3 = 98.67 \text{ ft}^2$$

$$R = \frac{A}{\text{wetted perimeter}} = \frac{98.67}{20 + 2 \times 3.3\sqrt{1 + 3^2}} = 2.414 \text{ ft}$$

By using the Chézy-Manning formula, Equation 8.22

$$Q = \frac{1.46}{n} A R^{2/3} S^{1/2}$$

$$S = \left[\frac{400 \times 0.025}{1.46 \times 98.67(2.414)^{2/3}}\right]^2 = 0.0015$$

From Equation 8.15,

$$S = S_e = \frac{H_1 - H_2}{L}$$

Hence the required channel length is

$$L = \frac{786 - 680}{0.0015} = 70{,}667 \text{ ft}$$

DESIGN OF STABLE CHANNELS

Channels built in erodible earth are susceptible to scouring as well as deposition of sediment carried away by the water. A recent field study done by the U.S. Bureau of Public Roads in Oklahoma resulted in useful practical information for design of stable earth channels. According to the USBR studies, scouring in an earth channel will not occur as long as the Froude number of the flow does not exceed 0.35.

Permissible maximum design velocities with regard to erosion of unlined earth channels in various soils were found to be equal

$$V = G y_n^{0.2} \qquad (8.28)$$

where y_n is the normal depth that must be substituted in ft, resulting in velocity di-mensions of ft/sec. For these units values of G erosion, coefficients for various soils are shown in Table 8-3. If the velocity is calculated to be too high to maintain a stable channel in an erodible soil, roadside ditches are built either with a concrete lining or with a flatter slope. The latter requires the construction of drop structures, small concrete weirs where the elevation difference is concentrated. The distance L between these drop structures is computed from the equation

$$L = \frac{100h}{S_1 - S_2} \qquad (8.29)$$

in which h is the height of the drop at the weir, S_1 is the slope of the original channel at which scour would occur, and S_2 is the new flattened slope, both in percent.

The permissible maximum channel slope for different soils is shown in Figure

Table 8-3 Erosion Coefficient of Various Soils

Soil Type	P.I., Plasticity Index	G, Erosion Coefficient
Sandy silt	1 to 6	2.5
Silt clay	6 to 10	2.5
Soft shale	—	3.2
Clay	10 or more	3.5
Soft sandstone	—	3.8

8.11 for various design discharges. This graph indicates that the selection of the proper design discharge is of great significance in designing a stable channel. Experience obtained by observing the performance of roadside ditches along interstate highways has shown that the scouring of the channels begins at the deepest points below the minimal permanent flow, the trickling discharge due to groundwater contributions between rains. For this reason the recent tendency of design is to provide a partial concrete lining limited to the bottom of the channel. This will prevent bottom scouring where grass cannot develop because of the permanent presence of water. On higher portions of the channel a strong grass growth is allowed to develop, which usually will prevent scouring during occasional high discharges.

Deposition of silt does not occur at Froude numbers 0.12 or larger unless the discharge is very large. In sandy silt and silt-clay soils, most susceptible for erosion and deposition, the design velocity must be kept above a minimum value,

$$V_{min} = 0.63 \, y_n^{0.64} \qquad (8.30)$$

where y_n, the normal depth, must be substituted in feet. The resulting minimum velocity is obtained by Equation 8.30 in ft/sec.

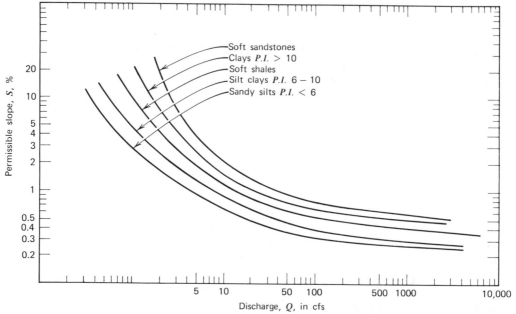

FIGURE 8.11 Permissible maximum channel slope in erodible soils.

EXAMPLE 8.17

The channel in Example 8.16 is to be dug in clay. Check if the channel will be susceptible to erosion or sediment deposition.

SOLUTION:

From Example 8.16 we have given

$$Q = 400 \text{ cfs}$$
$$n = 0.025$$
$$S = 0.0015$$
$$A = 98.67 \text{ ft}^2$$
$$R = 2.414 \text{ ft}$$
$$y_n = 3.3 \text{ ft}$$

$$v = \frac{Q}{A} = \frac{400}{98.67} = 4.05 \text{ fps}$$

1. Check velocity; $v_{min} = 0.63 \, y_n^{0.64}$, Equation 8.30.

$$v_{min} = 0.63 (3.3)^{0.64} = 1.35 \text{ fps} < 4.05 - OK$$

2. Check velocity; $v_{max} = G y_n^{0.2}$, Equation 8.28. From Table 8-3 for clay, $G = 3.5$. Therefore the allowable velocity

$$v_{max} = 3.5(3.3)^{0.2} = 4.44 \text{ fps} > 4.05 - OK$$

3. Check the permissible maximum channel slope. For clay from Figure 8.11 the maximum permissible slope of the channel is 0.0053, which is greater than $0.0015 - OK$.

Hence the channel is stable under the given conditions.

BACKWATER AND CHANNEL DELIVERY COMPUTATIONS

The limitations for use of the normal flow concept was made clear in the discussion of the delivery of channels in the previous parts of this chapter. Whenever there is a barrier in the downstream portion of the flow that causes the water to rise in order to continue along its path, the water surface cannot be assumed to be parallel with the bottom of the channel. In accordance with the gradual rise of the water depth toward the barrier the avail-able cross-sectional area enlarges, causing the velocity, hence the kinetic energy, to drop. Figure 8.12 shows two such cases frequently found in practice. One is the case of a dam that causes the flooding of the reservoir behind it, giving rise to the classical case of *backwater* flow. The other problem shown is conceptually identical to the first. This is also a backwater problem, with the exception that the water rise is caused by a flood in the recipient water body, such as a major river to which the channel being analyzed is a tributary. Such problems commonly occur, resulting in flooding at the lower

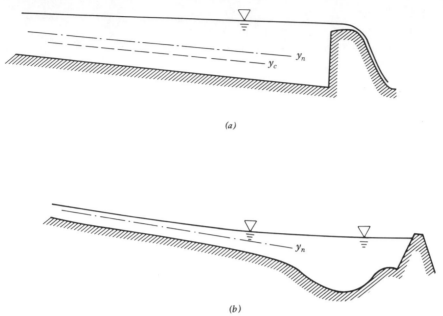

(a)

(b)

FIGURE 8.12   Examples of backwater flows. (a) Backwater above dam. (b) Backwater in tributary due to flood in main channel.

portions of tributary rivers and creeks. In order that proper plans be made for the clearing and deepening of such tributary channels or to determine the height requirements for projected levees around such tributaries, computations of backwater heights are necessary.

Note that backwater computations are different from flood flow computations. The second problem of the previous paragraph stems from a rise of the water level in a receiving waterbody downstream. In order for the channel to deliver its regular constant discharge a backwater will develop. Another case is when flooding that occurs in the channel is caused by a greatly increased discharge in the channel that the available channel cross section is unable to handle. Such unsteady flow problems are beyond the scope of this book.

While backwater computations are considered by some to be too involved, their importance in hydraulic practice is too great to exclude them from routine hydraulic work. Although the mathematical derivation of the necessary formulas are too complex to be reviewed here, the basic formula and the necessary steps for computation will be explained. For further information the reader is referred to the excellent books written on this subject, some of which are listed in the bibliography.

To show the computational technique of backwater flow consider the case of the tributary with a constant known discharge Q, known roughness value n, and a uniform channel bottom slope S. With the channel cross-sectional shape being known, all normal flow parameters, such as the normal depth y_n, as well as the critical depth y_c can be computed readily. The two hydraulic exponents, M and N, are also known. All of these are valid throughout the channel analyzed. To compute a backwater curve, its causing factor, the depth at the downstream end of the channel, y_0 should first be determined. This is the height to which the flow at the barrier must rise in order that the flow of the stated discharge Q may take place. In the case of a dam, the height y_0 at the dam must equal the height of the dam's spillway plus the thickness of the water nappe over the spillway. If the overflow conditions are not influenced by the downstream water elevation, the overflow height may be computed by the weir formula, Equation 9.9. Once y_0 is established, it will describe the end point of the backwater curve where the water is the deepest. The other end of the backwater curve approaches y_n toward the upstream in an asymptotic manner. For this reason the theoretical length of the backwater curve is infinite. For practical purposes the end can be established at the distance upstream where the water depth is, say, 1% greater than the normal depth, that is, where $y = 1.01 y_n$. Taking the average depth for the whole length of the backwater as

$$y_{ave} = \frac{y_0 + 1.01 y_n}{2}$$

and assuming that the cross-sectional shape of the channel does not change substantially over this length, the normal flow exponent N and the critical flow exponent M may be determined according to Equations 8.12 and 8.27, and the corresponding graphs.

Knowing N and M a new constant is to be formed that is

$$J = \frac{N}{N - M + 1} \qquad (8.31)$$

In order that the backwater curve be defined at several points by its height, a table is to be prepared as follows: Beginning with the greatest depth y_0 the depths to be located along the channel are determined arbitrarily. For example, if y_0 is 15 ft and y_n is 5 ft, the intermediate depths may be selected as $y_1 = 13$ ft, $y_2 = 11$ ft, $y_3 = 9$ ft, $y_4 = 7$ ft, $y_5 = 1.01 y_n$, or 5.05 ft. This will describe five yet unknown distances between the depths just selected. The backwater formula to be introduced below is set up so that it gives the distance between known water depths.

For each selected depth y_0 through y_5 express u_1, u_2, u_3 etc., such that

$$u = \frac{y}{y_n} \qquad (8.32)$$

Also express the values v_1, v_2, v_3, etc, such that

$$v = u^{N/J} \qquad (8.33)$$

where N is the average normal flow exponent and J is the result of Equation 8.31.

The computations can conveniently be done in a table of the form shown in Example 8.18. The values of $F(u, N)$ and $F(v, J)$ are to be found in Table 8-4. These are the so-called Bakhmeteff functions, corrected and extended by Van Te Chow. $F(v, J)$ is not shown in this table explicitly, but obtained by substituting v in place of u and J in place of N.

With the computational table completed through the F values the calculations proceed by using the backwater formula:

$$L = x_2 - x_1 = \frac{y_n}{S}\{(u_2 - u_1)$$
$$- [F(u_2, N) - F(u_1, N)]$$
$$+ B[F(v_2, J) - F(v_1, J)]\}$$
$$(8.34)$$

in which the subscripts 2 refer to the downstream end, and

$$B = \left(\frac{y_c}{y_n}\right)^M \frac{J}{N} \qquad (8.35)$$

a constant all along the channel.

Equation 8.34 gives the length of the channel between two adjacent depths listed as y_1, y_2, y_3, etc. For ease in computation the x values are computed first by

$$x = \frac{y_n}{S}[u - F(u, N) + BF(v, J)] \quad (8.36)$$

for each y depth. From these values the lengths are calculated. The distance of the location of each depth value measured from the downstream end is obtained by summing the consecutive L values.

After completing the table of computations the result may be plotted as y versus L along the channel. By connecting the resulting points by a smooth curve we obtain the backwater line. This information is essential in the design of flood control, drainage, and bridge elevations, for example—in all cases when the possibility of downstream influences are likely.

Note that the backwater curve refers to a particular discharge; for another discharge a different backwater curve will have to be computed.

EXAMPLE 8.18

A rectangular channel of 30 ft width carries a discharge of 450 cfs. The slope of the channel is 0.0017, its roughness factor is $n = 0.022$. The channel is tributary to a river in which the existing flood stage is 12 ft above the normal depth in the channel. Determine the resulting backwater curve in the tributary channel at four points along the channel (Figure E8.18).

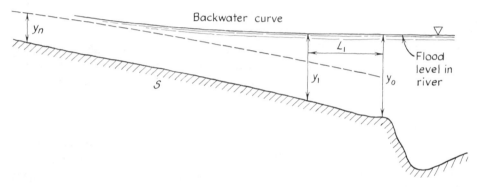

FIGURE E8.18

SOLUTION:

Given

$$Q = 450 \text{ cfs}$$
$$S = 0.0017$$
$$n = 0.022$$
$$b = 30 \text{ ft}$$

For a wide rectangular channel it may be assumed that the hydraulic radius equals the depth of the flow; hence, the normal depth can be computed by Equation 8.22 in which $A = b \cdot y_n$:

$$Q = \frac{1.49}{n} A y_n^{2/3} S^{1/2}$$

$$450 = \frac{1.49}{0.022} 30 y_n^{5/3} 0.0017^{1/2}$$

$$y_n = 2.74 \text{ ft}$$

The discharge per unit width is

$$q = \frac{Q}{B} = \frac{450}{30} = 15 \frac{\text{cfs}}{\text{ft}}$$

The critical depth may be computed by Equation 8.6.a

$$q_c = \sqrt{g \, y_c^3}$$

from which

$$y_c = \left(\frac{q^2}{g}\right)^{1/3} = \left(\frac{15^2}{32.2}\right)^{1/3} = 1.91 \text{ ft}$$

The upstream limit of the backwater curve will be selected as

$$y_4 = 1.05 y_n = 2.88 \text{ ft}$$

The downstream limit, at the entrance of the river is

$$y_0 = y_n + 12 = 14.74 \text{ ft}$$

The average depth is then

$$y_{\text{ave}} = \frac{14.74 + 2.88}{2} = 8.81 \text{ ft}$$

From

$$\frac{y_{\text{ave}}}{b} = \frac{8.81}{30} = 0.29$$

and by Figures 8.5 and 8.10 the hydraulic exponents are

$$M = 3.0 \qquad \text{and} \qquad N = 2.8$$

By Equation 8.31

$$J = \frac{2.8}{2.8 - 3 + 1} = 3.5$$

and

$$\frac{N}{J} = \frac{2.8}{3.5} = 0.8$$

From Equation 8.35 we obtain

$$B = \left(\frac{y_c}{y_n}\right)^M \cdot \frac{J}{N} = \left(\frac{1.91}{2.74}\right)^3 \cdot \frac{3.5}{2.8} = 0.42$$

The intermediate depths sought will be selected arbitrarily as round numbers between the limits of 2.88 and 14.74 ft, as follows:

$$y_0 = 14.74 \text{ ft}$$
$$y_1 = 12.0 \text{ ft}$$
$$y_2 = 8.0 \text{ ft}$$
$$y_3 = 4.0 \text{ ft}$$
$$y_4 = 2.88 \text{ ft}$$

The distances between the locations of these stage readings will be denoted as L_1, L_2, L_3, and L_4 and will be determined by Equation

8.34. The computations will be carried out in a tabular manner as shown in Table E8-18, using Equations 8.32 and 8.33. After determining the u and v values, the F functions are obtained for the sake of illustration as the nearest tabular values from Table 8-4. The tabulation is based on Equation 8.36.

Table E8-18

y ft	u	v	$F(u, N)$	$F(v, J)$	$u - F(u, N)$	$B \times F(v, J)$	$\dfrac{xS}{y_n}$	x	L	ΣL ft
14.74	5.37	3.84	0.27	0.05	5.10	0.035	5.135	8272	0	0
12.0	4.38	3.26	0.04	0.068	4.34	0.028	4.368	7037	1235	1235
8.0	2.92	2.35	0.082	0.125	2.83	0.049	2.859	4606	2431	3666
4.0	1.46	1.35	0.33	0.40	1.13	0.163	1.293	2083	2523	6189
2.88	1.06	1.04	0.95	1.08	0.11	0.45	0.56	902	1181	7370

The backwater formula, Equation 8.34, provides a solution for open channel flow problems under the most general conditions. Therefore its use is not limited to the determination of backwater-caused flooding only. As an example for the broad application of Equation 8.34, one may consider its use in the solution for the *delivery of channels*. If we refer to Figure 8.6 showing the delivery problem for a channel of L length under the conditions when the water level in the upper reservoir is kept constant and in the lower reservoir when the water level is varied, we see that the discharge variation due to the downstream water elevation may be determined. There are three points that are easy to obtain at this discharge-depth curve. These are the maximum at the point where the depth at the lower reservoir is critical, the normal discharge when the depths at both ends of the channel are equal, and the zero discharge when the lower water elevation is the same as the upper elevation. The discharge versus depth curve can easily and reliably be approximated by determining at least one more point on the curve.

Unfortunately there is no direct solution for a problem of this type. The backwater formula is written such that it results in a length L for known discharge

and its associated normal depth. In the delivery problem, on the other hand, the length L is known to begin with, and either y_0 or Q are the unknown variables. Because of the complexity of Equation 8.34, there is no way to rearrange it to give either of our unknowns in an explicit manner. Therefore, the only way to solve the problem is by using a trial and error approach.

Here is the trial and error procedure for solving the discharge delivery problem. First a discharge value is selected such that it is somewhat less than the value of the normal discharge. Selecting a discharge such that it is about 75% of the normal discharge is a reasonable start. Approximating the downstream depth to be about halfway between the normal depth and the level of the upper reservoir for this discharge may serve as a first trial value for y_0. Starting with these values the procedure shown for backwater computation may now be carried out, keeping in mind that y_0 is an arbitrarily selected value. For the selected discharge a new normal depth value should be calculated from Chézy's equation (or from Figures 8.8 or 8.9). Consequently the normal flow exponent N will be a new value. The critical flow exponent, M, may be calculated for the algebraic average of y_0 and

y_1. With these values Equations 8.31, 8.32, and 8.33 can now be determined. In this computation the upstream depth y_1 is the depth in the channel at the upper reservoir, which is now not equal to our newly calculated normal depth. After proper substitution into Equation 8.34, a trial value of L is obtained. Chances are this computed L will not be equal to the actual length of our channel. The computation now has to be repeated for a newly selected y_0, keeping the discharge Q the same as before. For the second trial y_0 should be selected such that if L_I computed was larger than L actual, the channel length, y_0 of the new trial should be smaller than its first assumed value.

With the new y_0 value the backwater computation procedure must be repeated resulting in a new L_{II} value.

Unless the two L values are significantly different from the actual length of the channel, further trial computations are not warranted. Instead, the proper value of y_0 associated with the selected Q may be determined by a linear approximation using the formula

$$y_0 = \left(\frac{y_{II} - y_I}{L_{II} - L_I}\right)(L_{actual} - L_I) + y_I \quad (8.37)$$

in which the values within the parenthesis are from our two trial solutions. The resulting y_0 value will be a reasonably correct approximation for the downstream depth for which our Q was selected to be 75% of our original Q_n. If we know now four points on the discharge delivery curve, it may be plotted giving a graphical result for all discharge-depth relationships under the conditions specified.

Table 8-4 Bakhmeteff's Functions (Reproduced by permission from Ven Te Chow: Integrating the equation of gradually varied flow, *Proceedings*, American Society of Civil Engineers, Vol. 81, Paper 838, pp. 1–32, November 1955.)

N / u	2.2	2.4	2.6	2.8	3.0	3.2	3.4	3.6	3.8	4.0
0.00	0.000	0.000	0.000	0.000	0.000	0.000	0.000	0.000	0.000	0.000
0.02	0.020	0.020	0.020	0.020	0.020	0.020	0.020	0.020	0.020	0.020
0.04	0.040	0.040	0.040	0.040	0.040	0.040	0.040	0.040	0.040	0.040
0.06	0.060	0.060	0.060	0.060	0.060	0.060	0.060	0.060	0.060	0.060
0.08	0.080	0.080	0.080	0.080	0.080	0.080	0.080	0.080	0.080	0.080
0.10	0.100	0.100	0.100	0.100	0.100	0.100	0.100	0.100	0.100	0.100
0.12	0.120	0.120	0.120	0.120	0.120	0.120	0.120	0.120	0.120	0.120
0.14	0.140	0.140	0.140	0.140	0.140	0.140	0.140	0.140	0.140	0.140
0.16	0.161	0.161	0.160	0.160	0.160	0.160	0.160	0.160	0.160	0.160
0.18	0.181	0.181	0.181	0.180	0.180	0.180	0.180	0.180	0.180	0.180
0.20	0.202	0.201	0.201	0.201	0.200	0.200	0.200	0.200	0.200	0.200
0.22	0.223	0.222	0.221	0.221	0.221	0.220	0.220	0.220	0.220	0.220
0.24	0.244	0.243	0.242	0.241	0.241	0.241	0.240	0.240	0.240	0.240
0.26	0.265	0.263	0.262	0.262	0.261	0.261	0.261	0.260	0.260	0.260
0.28	0.286	0.284	0.283	0.282	0.282	0.281	0.281	0.281	0.280	0.280
0.30	0.307	0.305	0.304	0.303	0.302	0.302	0.301	0.301	0.301	0.300
0.32	0.329	0.326	0.325	0.324	0.323	0.322	0.322	0.321	0.321	0.321
0.34	0.351	0.348	0.346	0.344	0.343	0.343	0.342	0.342	0.341	0.341
0.36	0.372	0.369	0.367	0.366	0.364	0.363	0.363	0.362	0.362	0.361
0.38	0.395	0.392	0.389	0.387	0.385	0.384	0.383	0.383	0.382	0.382

Table 8-4 *Cont'd*

N / u	2.2	2.4	2.6	2.8	3.0	3.2	3.4	3.6	3.8	4.0
0.40	0.418	0.414	0.411	0.408	0.407	0.405	0.404	0.403	0.403	0.402
0.42	0.442	0.437	0.433	0.430	0.428	0.426	0.425	0.424	0.423	0.423
0.44	0.465	0.460	0.456	0.452	0.450	0.448	0.446	0.445	0.444	0.443
0.46	0.489	0.483	0.479	0.475	0.472	0.470	0.468	0.466	0.465	0.464
0.48	0.514	0.507	0.502	0.497	0.494	0.492	0.489	0.488	0.486	0.485
0.50	0.539	0.531	0.525	0.521	0.517	0.514	0.511	0.509	0.508	0.506
0.52	0.565	0.557	0.550	0.544	0.540	0.536	0.534	0.531	0.529	0.528
0.54	0.592	0.582	0.574	0.568	0.563	0.559	0.556	0.554	0.551	0.550
0.56	0.619	0.608	0.599	0.593	0.587	0.583	0.579	0.576	0.574	0.572
0.58	0.648	0.635	0.626	0.618	0.612	0.607	0.603	0.599	0.596	0.594
0.60	0.676	0.663	0.653	0.644	0.637	0.631	0.627	0.623	0.620	0.617
0.61	0.691	0.678	0.667	0.657	0.650	0.644	0.639	0.635	0.631	0.628
0.62	0.706	0.692	0.680	0.671	0.663	0.657	0.651	0.647	0.643	0.640
0.63	0.722	0.707	0.694	0.684	0.676	0.669	0.664	0.659	0.655	0.652
0.64	0.738	0.722	0.709	0.698	0.690	0.683	0.677	0.672	0.667	0.664
0.65	0.754	0.737	0.724	0.712	0.703	0.696	0.689	0.684	0.680	0.676
0.66	0.771	0.753	0.738	0.727	0.717	0.709	0.703	0.697	0.692	0.688
0.67	0.787	0.769	0.754	0.742	0.731	0.723	0.716	0.710	0.705	0.701
0.68	0.804	0.785	0.769	0.757	0.746	0.737	0.729	0.723	0.718	0.713
0.69	0.822	0.804	0.785	0.772	0.761	0.751	0.743	0.737	0.731	0.726
0.70	0.840	0.819	0.802	0.787	0.776	0.766	0.757	0.750	0.744	0.739
0.71	0.858	0.836	0.819	0.804	0.791	0.781	0.772	0.764	0.758	0.752
0.72	0.878	0.855	0.836	0.820	0.807	0.796	0.786	0.779	0.772	0.766
0.73	0.898	0.874	0.854	0.837	0.823	0.811	0.802	0.793	0.786	0.780
0.74	0.918	0.892	0.868	0.854	0.840	0.827	0.817	0.808	0.800	0.794
0.75	0.940	0.913	0.890	0.872	0.857	0.844	0.833	0.823	0.815	0.808
0.76	0.961	0.933	0.909	0.890	0.874	0.861	0.849	0.839	0.830	0.823
0.77	0.985	0.954	0.930	0.909	0.892	0.878	0.866	0.855	0.846	0.838
0.78	1.007	0.976	0.950	0.929	0.911	0.896	0.883	0.872	0.862	0.854
0.79	1.031	0.998	0.971	0.949	0.930	0.914	0.901	0.889	0.879	0.870
0.80	1.056	1.022	0.994	0.970	0.950	0.934	0.919	0.907	0.896	0.887
0.81	1.083	1.046	1.017	0.992	0.971	0.954	0.938	0.925	0.914	0.904
0.82	1.110	1.072	1.044	1.015	0.993	0.974	0.958	0.945	0.932	0.922
0.83	1.139	1.099	1.067	1.039	1.016	0.996	0.979	0.965	0.952	0.940
0.84	1.171	1.129	1.094	1.064	1.040	1.019	1.001	0.985	0.972	0.960
0.85	1.201	1.157	1.121	1.091	1.065	1.043	1.024	1.007	0.993	0.980
0.86	1.238	1.192	1.153	1.119	1.092	1.068	1.048	1.031	1.015	1.002
0.87	1.272	1.223	1.182	1.149	1.120	1.095	1.074	1.055	1.039	1.025
0.88	1.314	1.262	1.228	1.181	1.151	1.124	1.101	1.081	1.064	1.049
0.89	1.357	1.302	1.255	1.216	1.183	1.155	1.131	1.110	1.091	1.075

Table 8-4 *Cont'd*

N \ u	2.2	2.4	2.6	2.8	3.0	3.2	3.4	3.6	3.8	4.0
0.90	1.401	1.343	1.294	1.253	1.218	1.189	1.163	1.140	1.120	1.103
0.91	1.452	1.389	1.338	1.294	1.257	1.225	1.197	1.173	1.152	1.133
0.92	1.505	1.438	1.351	1.340	1.300	1.266	1.236	1.210	1.187	1.166
0.93	1.564	1.493	1.435	1.391	1.348	1.311	1.279	1.251	1.226	1.204
0.94	1.645	1.568	1.504	1.449	1.403	1.363	1.328	1.297	1.270	1.246
0.950	1.737	1.652	1.582	1.518	1.467	1.423	1.385	1.352	1.322	1.296
0.960	1.833	1.741	1.665	1.601	1.545	1.497	1.454	1.417	1.385	1.355
0.970	1.969	1.866	1.780	1.707	1.644	1.590	1.543	1.501	1.464	1.431
0.975	2.055	1.945	1.853	1.773	1.707	1.649	1.598	1.554	1.514	1.479
0.980	2.164	2.045	1.946	1.855	1.783	1.720	1.666	1.617	1.575	1.536
0.985	2.294	2.165	2.056	1.959	1.880	1.812	1.752	1.699	1.652	1.610
0.990	2.477	2.333	2.212	2.106	2.017	1.910	1.873	1.814	1.761	1.714
0.995	2.792	2.621	2.478	2.355	2.250	2.159	2.079	2.008	1.945	1.889
0.999	3.523	3.292	3.097	2.931	2.788	2.663	2.554	2.457	2.370	2.293
1.000	∞	∞	∞	∞	∞	∞	∞	∞	∞	∞
1.001	3.317	2.931	2.640	2.399	2.184	2.008	1.856	1.725	1.610	1.508
1.005	2.587	2.266	2.022	1.818	1.649	1.506	1.384	1.279	1.188	1.107
1.010	2.273	1.977	1.757	1.572	1.419	1.291	1.182	1.089	1.007	0.936
1.015	2.090	1.807	1.602	1.428	1.286	1.166	1.065	0.978	0.902	0.836
1.020	1.961	1.711	1.493	1.327	1.191	1.078	0.982	0.900	0.828	0.766
1.03	1.779	1.531	1.340	1.186	1.060	0.955	0.866	0.790	0.725	0.668
1.04	1.651	1.410	1.232	1.086	0.967	0.868	0.785	0.714	0.653	0.600
1.05	1.552	1.334	1.150	1.010	0.896	0.802	0.723	0.656	0.598	0.548
1.06	1.472	1.250	1.082	0.948	0.838	0.748	0.672	0.608	0.553	0.506
1.07	1.404	1.195	1.026	0.896	0.790	0.703	0.630	0.569	0.516	0.471
1.08	1.346	1.139	0.978	0.851	0.749	0.665	0.595	0.535	0.485	0.441
1.09	1.295	1.089	0.935	0.812	0.713	0.631	0.563	0.506	0.457	0.415
1.10	1.250	1.050	0.897	0.777	0.681	0.601	0.536	0.480	0.433	0.392
1.11	1.209	1.014	0.864	0.746	0.652	0.575	0.511	0.457	0.411	0.372
1.12	1.172	0.981	0.833	0.718	0.626	0.551	0.488	0.436	0.392	0.354
1.13	1.138	0.950	0.805	0.692	0.602	0.520	0.468	0.417	0.371	0.337
1.14	1.107	0.921	0.780	0.669	0.581	0.509	0.450	0.400	0.358	0.322
1.15	1.078	0.892	0.756	0.647	0.561	0.490	0.432	0.384	0.343	0.308
1.16	1.052	0.870	0.734	0.627	0.542	0.473	0.417	0.369	0.329	0.295
1.17	1.027	0.850	0.713	0.608	0.525	0.458	0.402	0.356	0.317	0.283
1.18	1.003	0.825	0.694	0.591	0.509	0.443	0.388	0.343	0.305	0.272
1.19	0.981	0.810	0.676	0.574	0.494	0.429	0.375	0.331	0.294	0.262
1.20	0.960	0.787	0.659	0.559	0.480	0.416	0.363	0.320	0.283	0.252
1.22	0.922	0.755	0.628	0.531	0.454	0.392	0.344	0.299	0.264	0.235
1.24	0.887	0.725	0.600	0.505	0.431	0.371	0.322	0.281	0.248	0.219

Table 8-4 *Cont'd*

u \ N	2.2	2.4	2.6	2.8	3.0	3.2	3.4	3.6	3.8	4.0
1.26	0.855	0.692	0.574	0.482	0.410	0.351	0.304	0.265	0.233	0.205
1.28	0.827	0.666	0.551	0.461	0.391	0.334	0.288	0.250	0.219	0.193
1.30	0.800	0.644	0.530	0.442	0.373	0.318	0.274	0.237	0.207	0.181
1.32	0.775	0.625	0.510	0.424	0.357	0.304	0.260	0.225	0.196	0.171
1.34	0.752	0.605	0.492	0.408	0.342	0.290	0.248	0.214	0.185	0.162
1.36	0.731	0.588	0.475	0.393	0.329	0.278	0.237	0.204	0.176	0.153
1.38	0.711	0.567	0.459	0.378	0.316	0.266	0.226	0.194	0.167	0.145
1.40	0.692	0.548	0.444	0.365	0.304	0.256	0.217	0.185	0.159	0.138
1.42	0.674	0.533	0.431	0.353	0.293	0.246	0.208	0.177	0.152	0.131
1.44	0.658	0.517	0.417	0.341	0.282	0.236	0.199	0.169	0.145	0.125
1.46	0.642	0.505	0.405	0.330	0.273	0.227	0.191	0.162	0.139	0.119
1.48	0.627	0.493	0.394	0.320	0.263	0.219	0.184	0.156	0.133	0.113
1.50	0.613	0.480	0.383	0.310	0.255	0.211	0.177	0.149	0.127	0.108
1.55	0.580	0.451	0.358	0.288	0.235	0.194	0.161	0.135	0.114	0.097
1.60	0.551	0.425	0.335	0.269	0.218	0.179	0.148	0.123	0.103	0.087
1.65	0.525	0.402	0.316	0.251	0.203	0.165	0.136	0.113	0.094	0.079
1.70	0.501	0.381	0.298	0.236	0.189	0.153	0.125	0.103	0.086	0.072
1.75	0.480	0.362	0.282	0.222	0.177	0.143	0.116	0.095	0.079	0.065
1.80	0.460	0.349	0.267	0.209	0.166	0.133	0.108	0.088	0.072	0.060
1.85	0.442	0.332	0.254	0.198	0.156	0.125	0.100	0.082	0.067	0.055
1.90	0.425	0.315	0.242	0.188	0.147	0.117	0.094	0.076	0.062	0.050
1.95	0.409	0.304	0.231	0.178	0.139	0.110	0.088	0.070	0.057	0.046
2.00	0.395	0.292	0.221	0.169	0.132	0.104	0.082	0.066	0.053	0.043
2.10	0.369	0.273	0.202	0.154	0.119	0.092	0.073	0.058	0.046	0.037
2.20	0.346	0.253	0.186	0.141	0.107	0.083	0.065	0.051	0.040	0.032
2.3	0.326	0.235	0.173	0.129	0.098	0.075	0.058	0.045	0.035	0.028
2.4	0.308	0.220	0.160	0.119	0.089	0.068	0.052	0.040	0.031	0.024
2.5	0.292	0.207	0.150	0.110	0.082	0.062	0.047	0.036	0.028	0.022
2.6	0.277	0.197	0.140	0.102	0.076	0.057	0.043	0.033	0.025	0.019
2.7	0.264	0.188	0.131	0.095	0.070	0.052	0.039	0.029	0.022	0.017
2.8	0.252	0.176	0.124	0.089	0.065	0.048	0.036	0.027	0.020	0.015
2.9	0.241	0.166	0.117	0.083	0.060	0.044	0.033	0.024	0.018	0.014
3.0	0.230	0.159	0.110	0.078	0.056	0.041	0.030	0.022	0.017	0.012
3.5	0.190	0.126	0.085	0.059	0.041	0.029	0.021	0.015	0.011	0.008
4.0	0.161	0.104	0.069	0.046	0.031	0.022	0.015	0.010	0.007	0.005
4.5	0.139	0.087	0.057	0.037	0.025	0.017	0.011	0.008	0.005	0.004
5.0	0.122	0.076	0.048	0.031	0.020	0.013	0.009	0.006	0.004	0.003
6.0	0.098	0.060	0.036	0.022	0.014	0.009	0.006	0.004	0.002	0.002
7.0	0.081	0.048	0.028	0.017	0.010	0.006	0.004	0.002	0.002	0.001
8.0	0.069	0.040	0.022	0.013	0.008	0.005	0.003	0.002	0.001	0.001
9.0	0.060	0.034	0.019	0.011	0.006	0.004	0.002	0.001	0.001	0.000
10.0	0.053	0.028	0.016	0.009	0.005	0.003	0.002	0.001	0.001	0.000
20.0	0.023	0.018	0.011	0.006	0.002	0.001	0.001	0.000	0.000	0.000

Table 8-4 *Cont'd*

N u	4.2	4.6	5.0	5.4	5.8	6.2	6.6	7.0	7.4	7.8
0.00	0.000	0.000	0.000	0.000	0.000	0.000	0.000	0.000	0.000	0.000
0.02	0.020	0.020	0.020	0.020	0.020	0.020	0.020	0.020	0.020	0.020
0.04	0.040	0.040	0.040	0.040	0.040	0.040	0.040	0.040	0.040	0.040
0.06	0.060	0.060	0.060	0.060	0.060	0.060	0.060	0.060	0.060	0.060
0.08	0.080	0.080	0.080	0.080	0.080	0.080	0.080	0.080	0.080	0.080
0.10	0.100	0.100	0.100	0.100	0.100	0.100	0.100	0.100	0.100	0.100
0.12	0.120	0.120	0.120	0.120	0.120	0.120	0.120	0.120	0.120	0.120
0.14	0.140	0.140	0.140	0.140	0.140	0.140	0.140	0.140	0.140	0.140
0.16	0.160	0.160	0.160	0.160	0.160	0.160	0.160	0.160	0.160	0.160
0.18	0.180	0.180	0.180	0.180	0.180	0.180	0.180	0.180	0.180	0.180
0.20	0.200	0.200	0.200	0.200	0.200	0.200	0.200	0.200	0.200	0.200
0.22	0.220	0.220	0.220	0.220	0.220	0.220	0.220	0.220	0.220	0.220
0.24	0.240	0.240	0.240	0.240	0.240	0.240	0.240	0.240	0.240	0.240
0.26	0.260	0.260	0.260	0.260	0.260	0.260	0.260	0.260	0.260	0.260
0.28	0.280	0.280	0.280	0.280	0.280	0.280	0.280	0.280	0.280	0.280
0.30	0.300	0.300	0.300	0.300	0.300	0.300	0.300	0.300	0.300	0.300
0.32	0.321	0.320	0.320	0.320	0.320	0.320	0.320	0.320	0.320	0.320
0.34	0.341	0.340	0.340	0.340	0.340	0.340	0.340	0.340	0.340	0.340
0.36	0.361	0.361	0.360	0.360	0.360	0.360	0.360	0.360	0.360	0.360
0.38	0.381	0.381	0.381	0.380	0.380	0.380	0.380	0.380	0.380	0.380
0.40	0.402	0.401	0.401	0.400	0.400	0.400	0.400	0.400	0.400	0.400
0.42	0.422	0.421	0.421	0.421	0.420	0.420	0.420	0.420	0.420	0.420
0.44	0.443	0.442	0.441	0.441	0.441	0.441	0.440	0.440	0.440	0.440
0.46	0.463	0.462	0.462	0.461	0.461	0.461	0.460	0.460	0.460	0.460
0.48	0.484	0.483	0.482	0.481	0.481	0.481	0.480	0.480	0.480	0.480
0.50	0.505	0.504	0.503	0.502	0.501	0.501	0.501	0.500	0.500	0.500
0.52	0.527	0.525	0.523	0.522	0.522	0.521	0.521	0.521	0.520	0.520
0.54	0.548	0.546	0.544	0.543	0.542	0.542	0.541	0.541	0.541	0.541
0.56	0.570	0.567	0.565	0.564	0.563	0.562	0.562	0.561	0.561	0.561
0.58	0.592	0.589	0.587	0.585	0.583	0.583	0.582	0.582	0.581	0.581
0.60	0.614	0.611	0.608	0.606	0.605	0.604	0.603	0.602	0.602	0.601
0.61	0.626	0.622	0.619	0.617	0.615	0.614	0.613	0.612	0.612	0.611
0.62	0.637	0.633	0.630	0.628	0.626	0.625	0.624	0.623	0.622	0.622
0.63	0.649	0.644	0.641	0.638	0.636	0.635	0.634	0.633	0.632	0.632
0.64	0.661	0.656	0.652	0.649	0.647	0.646	0.645	0.644	0.643	0.642
0.65	0.673	0.667	0.663	0.660	0.658	0.656	0.655	0.654	0.653	0.653
0.66	0.685	0.679	0.675	0.672	0.669	0.667	0.666	0.665	0.664	0.663
0.67	0.697	0.691	0.686	0.683	0.680	0.678	0.676	0.675	0.674	0.673
0.68	0.709	0.703	0.698	0.694	0.691	0.689	0.687	0.686	0.685	0.684
0.69	0.722	0.715	0.710	0.706	0.703	0.700	0.698	0.696	0.695	0.694

Table 8-4 *Cont'd*

N \ u	4.2	4.6	5.0	5.4	5.8	6.2	6.6	7.0	7.4	7.8
0.70	0.735	0.727	0.722	0.717	0.714	0.712	0.710	0.708	0.706	0.705
0.71	0.748	0.740	0.734	0.729	0.726	0.723	0.721	0.719	0.717	0.716
0.72	0.761	0.752	0.746	0.741	0.737	0.734	0.732	0.730	0.728	0.727
0.73	0.774	0.765	0.759	0.753	0.749	0.746	0.743	0.741	0.739	0.737
0.74	0.788	0.779	0.771	0.766	0.761	0.757	0.754	0.752	0.750	0.748
0.75	0.802	0.792	0.784	0.778	0.773	0.769	0.766	0.763	0.761	0.759
0.76	0.817	0.806	0.798	0.791	0.786	0.782	0.778	0.775	0.773	0.771
0.77	0.831	0.820	0.811	0.804	0.798	0.794	0.790	0.787	0.784	0.782
0.78	0.847	0.834	0.825	0.817	0.811	0.806	0.802	0.799	0.796	0.794
0.79	0.862	0.849	0.839	0.831	0.824	0.819	0.815	0.811	0.808	0.805
0.80	0.878	0.865	0.854	0.845	0.838	0.832	0.828	0.823	0.820	0.818
0.81	0.895	0.881	0.869	0.860	0.852	0.846	0.841	0.836	0.833	0.830
0.82	0.913	0.897	0.885	0.875	0.866	0.860	0.854	0.850	0.846	0.842
0.83	0.931	0.914	0.901	0.890	0.881	0.874	0.868	0.863	0.859	0.855
0.84	0.949	0.932	0.918	0.906	0.897	0.889	0.882	0.877	0.872	0.868
0.85	0.969	0.950	0.935	0.923	0.912	0.905	0.898	0.891	0.887	0.882
0.86	0.990	0.970	0.954	0.940	0.930	0.921	0.913	0.906	0.901	0.896
0.87	1.012	0.990	0.973	0.959	0.947	0.937	0.929	0.922	0.916	0.911
0.88	1.035	1.012	0.994	0.978	0.966	0.955	0.946	0.938	0.932	0.927
0.89	1.060	1.035	1.015	0.999	0.986	0.974	0.964	0.956	0.949	0.943
0.90	1.087	1.060	1.039	1.021	1.007	0.994	0.984	0.974	0.967	0.960
0.91	1.116	1.088	1.064	1.045	1.029	1.016	1.003	0.995	0.986	0.979
0.92	1.148	1.117	1.092	1.072	1.054	1.039	1.027	1.016	1.006	0.999
0.93	1.184	1.151	1.123	1.101	1.081	1.065	1.050	1.040	1.029	1.021
0.94	1.225	1.188	1.158	1.134	1.113	1.095	1.080	1.066	1.054	1.044
0.950	1.272	1.232	1.199	1.172	1.148	1.128	1.111	1.097	1.084	1.073
0.960	1.329	1.285	1.248	1.217	1.188	1.167	1.149	1.133	1.119	1.106
0.970	1.402	1.351	1.310	1.275	1.246	1.219	1.197	1.179	1.162	1.148
0.975	1.447	1.393	1.348	1.311	1.280	1.250	1.227	1.207	1.190	1.173
0.980	1.502	1.443	1.395	1.354	1.339	1.288	1.262	1.241	1.221	1.204
0.985	1.573	1.508	1.454	1.409	1.372	1.337	1.309	1.284	1.263	1.243
0.990	1.671	1.598	1.537	1.487	1.444	1.404	1.373	1.344	1.319	1.297
0.995	1.838	1.751	1.678	1.617	1.565	1.519	1.479	1.451	1.419	1.388
0.999	2.223	2.102	2.002	1.917	1.845	1.780	1.725	1.678	1.635	1.596
1.000	∞	∞	∞	∞	∞	∞	∞	∞	∞	∞
1.001	1.417	1.264	1.138	1.033	0.951	0.870	0.803	0.746	0.697	0.651
1.005	1.036	0.915	0.817	0.737	0.669	0.612	0.553	0.526	0.481	0.447
1.010	0.873	0.766	0.681	0.610	0.551	0.502	0.459	0.422	0.389	0.360
1.015	0.778	0.680	0.602	0.537	0.483	0.440	0.399	0.366	0.336	0.310
1.02	0.711	0.620	0.546	0.486	0.436	0.394	0.358	0.327	0.300	0.276

Table 8-4 *Cont'd*

N u	4.2	4.6	5.0	5.4	5.8	6.2	6.6	7.0	7.4	7.8
1.03	0.618	0.535	0.469	0.415	0.370	0.333	0.300	0.272	0.299	0.228
1.04	0.554	0.477	0.415	0.365	0.324	0.290	0.262	0.236	0.211	0.195
1.05	0.504	0.432	0.374	0.328	0.289	0.259	0.231	0.208	0.180	0.174
1.06	0.464	0.396	0.342	0.298	0.262	0.233	0.209	0.187	0.170	0.154
1.07	0.431	0.366	0.315	0.273	0.239	0.212	0.191	0.168	0.154	0.136
1.08	0.403	0.341	0.292	0.252	0.220	0.194	0.172	0.153	0.137	0.123
1.09	0.379	0.319	0.272	0.234	0.204	0.179	0.158	0.140	0.125	0.112
1.10	0.357	0.299	0.254	0.218	0.189	0.165	0.146	0.129	0.114	0.102
1.11	0.338	0.282	0.239	0.201	0.176	0.154	0.135	0.119	0.105	0.094
1.12	0.321	0.267	0.225	0.192	0.165	0.143	0.125	0.110	0.097	0.080
1.13	0.305	0.253	0.212	0.181	0.155	0.135	0.117	0.102	0.099	0.080
1.14	0.291	0.240	0.201	0.170	0.146	0.126	0.109	0.095	0.084	0.074
1.15	0.278	0.229	0.191	0.161	0.137	0.118	0.102	0.089	0.078	0.068
1.16	0.266	0.218	0.181	0.153	0.130	0.111	0.096	0.084	0.072	0.064
1.17	0.255	0.208	0.173	0.145	0.123	0.105	0.090	0.078	0.068	0.060
1.18	0.244	0.199	0.165	0.138	0.116	0.099	0.085	0.073	0.063	0.055
1.19	0.235	0.191	0.157	0.131	0.110	0.094	0.080	0.068	0.059	0.051
1.20	0.226	0.183	0.150	0.215	0.105	0.088	0.076	0.064	0.056	0.048
1.22	0.209	0.168	0.138	0.114	0.095	0.080	0.068	0.057	0.049	0.042
1.24	0.195	0.156	0.127	0.104	0.086	0.072	0.060	0.051	0.044	0.038
1.26	0.182	0.145	0.117	0.095	0.079	0.065	0.055	0.046	0.039	0.033
1.28	0.170	0.135	0.108	0.088	0.072	0.060	0.050	0.041	0.035	0.030
1.30	0.160	0.126	0.100	0.081	0.066	0.054	0.045	0.037	0.031	0.026
1.32	0.150	0.118	0.093	0.075	0.061	0.050	0.041	0.034	0.028	0.024
1.34	0.142	0.110	0.087	0.069	0.056	0.045	0.037	0.030	0.025	0.021
1.36	0.134	0.103	0.081	0.064	0.052	0.042	0.034	0.028	0.023	0.019
1.38	0.127	0.097	0.076	0.060	0.048	0.038	0.032	0.026	0.021	0.017
1.40	0.120	0.092	0.074	0.056	0.044	0.036	0.028	0.023	0.019	0.016
1.42	0.114	0.087	0.067	0.052	0.041	0.033	0.026	0.021	0.017	0.014
1.44	0.108	0.082	0.063	0.049	0.038	0.030	0.024	0.019	0.016	0.013
1.46	0.103	0.077	0.059	0.046	0.036	0.028	0.022	0.018	0.014	0.012
1.48	0.098	0.073	0.056	0.043	0.033	0.026	0.021	0.017	0.013	0.010
1.50	0.093	0.069	0.053	0.040	0.031	0.024	0.020	0.015	0.012	0.009
1.55	0.083	0.061	0.046	0.035	0.026	0.020	0.016	0.012	0.010	0.008
1.60	0.074	0.054	0.040	0.030	0.023	0.017	0.013	0.010	0.008	0.006
1.65	0.067	0.048	0.035	0.026	0.019	0.014	0.011	0.008	0.006	0.005
1.70	0.060	0.043	0.031	0.023	0.016	0.012	0.009	0.007	0.005	0.004
1.75	0.054	0.038	0.027	0.020	0.014	0.010	0.008	0.006	0.004	0.003
1.80	0.019	0.034	0.024	0.017	0.012	0.009	0.007	0.005	0.004	0.003
1.85	0.045	0.031	0.022	0.015	0.011	0.008	0.006	0.004	0.003	0.002

Table 8-4 *Cont'd*

N \ u	4.2	4.6	5.0	5.4	5.8	6.2	6.6	7.0	7.4	7.8
1.90	0.041	0.028	0.020	0.014	0.010	0.007	0.005	0.004	0.003	0.002
1.95	0.038	0.026	0.018	0.012	0.008	0.006	0.004	0.003	0.002	0.002
2.00	0.035	0.023	0.016	0.011	0.007	0.005	0.004	0.003	0.002	0.001
2.10	0.030	0.019	0.013	0.009	0.006	0.004	0.003	0.002	0.001	0.001
2.20	0.025	0.016	0.011	0.007	0.005	0.004	0.002	0.001	0.001	0.001
2.3	0.022	0.014	0.009	0.006	0.004	0.003	0.002	0.001	0.001	0.001
2.4	0.019	0.012	0.008	0.005	0.003	0.002	0.001	0.001	0.001	0.001
2.5	0.017	0.010	0.006	0.004	0.003	0.002	0.001	0.001	0.000	0.000
2.6	0.015	0.009	0.005	0.003	0.002	0.001	0.001	0.001	0.000	0.000
2.7	0.013	0.008	0.005	0.003	0.002	0.001	0.001	0.000	0.000	0.000
2.8	0.012	0.007	0.004	0.002	0.001	0.001	0.001	0.000	0.000	0.000
2.9	0.010	0.006	0.004	0.002	0.001	0.001	0.000	0.000	0.000	0.000
3.0	0.009	0.005	0.003	0.002	0.001	0.001	0.000	0.000	0.000	0.000
3.5	0.006	0.003	0.002	0.001	0.001	0.000	0.000	0.000	0.000	0.000
4.0	0.004	0.002	0.001	0.000	0.000	0.000	0.000	0.000	0.000	0.000
4.5	0.003	0.001	0.001	0.000	0.000	0.000	0.000	0.000	0.000	0.000
5.0	0.002	0.001	0.000	0.000	0.000	0.000	0.000	0.000	0.000	0.000
6.0	0.001	0.000	0.000	0.000	0.000	0.000	0.000	0.000	0.000	0.000
7.0	0.001	0.000	0.000	0.000	0.000	0.000	0.000	0.000	0.000	0.000
8.0	0.000	0.000	0.000	0.000	0.000	0.000	0.000	0.000	0.000	0.000
9.0	0.000	0.000	0.000	0.000	0.000	0.000	0.000	0.000	0.000	0.000
10.0	0.000	0.000	0.000	0.000	0.000	0.000	0.000	0.000	0.000	0.000
20.0	0.000	0.000	0.000	0.000	0.000	0.000	0.000	0.000	0.000	0.000

N \ u	8.2	8.6	9.0	9.4	9.8
0.00	0.000	0.000	0.000	0.000	0.000
0.02	0.020	0.020	0.020	0.020	0.020
0.04	0.040	0.040	0.040	0.040	0.040
0.06	0.060	0.060	0.060	0.060	0.060
0.08	0.080	0.080	0.080	0.080	0.080
0.10	0.100	0.100	0.100	0.100	0.100
0.12	0.120	0.120	0.120	0.120	0.120
0.14	0.140	0.140	0.140	0.140	0.140
0.16	0.160	0.160	0.160	0.160	0.160
0.18	0.180	0.180	0.180	0.180	0.180
0.20	0.200	0.200	0.200	0.200	0.200
0.22	0.220	0.220	0.220	0.220	0.220
0.24	0.240	0.240	0.240	0.240	0.240
0.26	0.260	0.260	0.260	0.260	0.260
0.28	0.280	0.280	0.280	0.280	0.280

Table 8-4 *Cont'd*

N u	8.2	8.6	9.0	9.4	9.8
0.30	0.300	0.300	0.300	0.300	0.300
0.32	0.320	0.320	0.320	0.320	0.320
0.34	0.340	0.340	0.340	0.340	0.340
0.36	0.360	0.360	0.360	0.360	0.360
0.38	0.380	0.380	0.380	0.380	0.380
0.40	0.400	0.400	0.400	0.400	0.400
0.42	0.420	0.420	0.420	0.420	0.420
0.44	0.440	0.440	0.440	0.440	0.440
0.46	0.460	0.460	0.460	0.460	0.460
0.48	0.480	0.480	0.480	0.480	0.480
0.50	0.500	0.500	0.500	0.500	0.500
0.52	0.520	0.520	0.520	0.520	0.520
0.54	0.540	0.540	0.540	0.540	0.540
0.56	0.561	0.560	0.560	0.560	0.560
0.58	0.581	0.581	0.580	0.580	0.580
0.60	0.601	0.601	0.601	0.600	0.600
0.61	0.611	0.611	0.611	0.611	0.610
0.62	0.621	0.621	0.621	0.621	0.621
0.63	0.632	0.631	0.631	0.631	0.631
0.64	0.642	0.641	0.641	0.641	0.641
0.65	0.652	0.652	0.651	0.651	0.651
0.66	0.662	0.662	0.662	0.661	0.661
0.67	0.673	0.672	0.672	0.672	0.671
0.68	0.683	0.683	0.682	0.682	0.681
0.69	0.694	0.693	0.692	0.692	0.692
0.70	0.704	0.704	0.703	0.702	0.702
0.71	0.715	0.714	0.713	0.713	0.712
0.72	0.726	0.725	0.724	0.723	0.723
0.73	0.736	0.735	0.734	0.734	0.733
0.74	0.747	0.746	0.745	0.744	9.744
0.75	0.758	0.757	0.756	0.755	0.754
0.76	0.769	0.768	0.767	0.766	0.765
0.77	0.780	0.779	0.778	0.777	0.776
0.78	0.792	0.790	0.789	0.788	0.787
0.79	0.804	0.802	0.800	0.799	0.798
0.80	0.815	0.813	0.811	0.810	0.809
0.81	0.827	0.825	0.823	0.822	0.820
0.82	0.839	0.837	0.835	0.833	0.831
0.83	0.852	0.849	0.847	0.845	0.844
0.84	0.865	0.862	0.860	0.858	0.856

Table 8-4   *Cont'd*

N u	8.2	8.6	9.0	9.4	9.8
0.85	0.878	0.875	0.873	0.870	0.868
0.86	0.892	0.889	0.886	0.883	0.881
0.87	0.907	0.903	0.900	0.897	0.894
0.88	0.921	0.918	0.914	0.911	0.908
0.89	0.937	0.933	0.929	0.925	0.922
0.90	0.954	0.949	0.944	0.940	0.937
0.91	0.972	0.967	0.961	0.957	0.953
0.92	0.991	0.986	0.980	0.975	0.970
0.93	1.012	1.006	0.999	0.994	0.989
0.94	1.036	1.029	1.022	1.016	1.010
0.950	1.062	1.055	1.047	1.040	1.033
0.960	1.097	1.085	1.074	1.063	1.053
0.970	1.136	1.124	1.112	1.100	1.087
0.975	1.157	1.147	1.134	1.122	1.108
0.980	1.187	1.175	1.160	1.150	1.132
0.985	1.224	1.210	1.196	1.183	1.165
0.990	1.275	1.260	1.243	1.228	1.208
0.995	1.363	1.342	1.320	1.302	1.280
0.999	1.560	1.530	1.500	1.476	1.447
1.000	∞	∞	∞	∞	∞
1.001	0.614	0.577	0.546	0.519	0.494
1.005	0.420	0.391	0.368	0.350	0.331
1.010	0.337	0.313	0.294	0.278	0.262
1.015	0.289	0.269	0.255	0.237	0.223
1.020	0.257	0.237	0.221	0.209	0.196
1.03	0.212	0.195	0.181	0.170	0.159
1.04	0.173	0.165	0.152	0.143	0.134
1.05	0.158	0.143	0.132	0.124	0.115
1.06	0.140	0.127	0.116	0.106	0.098
1.07	0.123	0.112	0.102	0.094	0.086
1.08	0.111	0.101	0.092	0.084	0.077
1.09	0.101	0.091	0.082	0.075	0.069
1.10	0.092	0.083	0.074	0.067	0.062
1.11	0.084	0.075	0.067	0.060	0.055
1.12	0.077	0.069	0.062	0.055	0.050
1.13	0.071	0.063	0.056	0.050	0.045
1.14	0.065	0.058	0.052	0.046	0.041
1.15	0.061	0.054	0.048	0.043	0.038
1.16	0.056	0.050	0.045	0.040	0.035
1.17	0.052	0.046	0.041	0.036	0.032

Table 8-4 *Cont'd*

u \ N	8.2	8.6	9.0	9.4	9.8
1.18	0.048	0.042	0.037	0.033	0.029
1.19	0.045	0.039	0.034	0.030	0.027
1.20	0.043	0.037	0.032	0.028	0.025
1.22	0.037	0.032	0.028	0.024	0.021
1.24	0.032	0.028	0.024	0.021	0.018
1.26	0.028	0.024	0.021	0.018	0.016
1.28	0.025	0.021	0.018	0.016	0.014
1.30	0.022	0.019	0.016	0.014	0.012
1.32	0.020	0.017	0.014	0.012	0.010
1.34	0.018	0.015	0.012	0.010	0.009
1.36	0.016	0.013	0.011	0.009	0.008
1.38	0.014	0.012	0.010	0.008	0.007
1.40	0.013	0.011	0.009	0.007	0.006
1.42	0.011	0.009	0.008	0.006	0.005
1.44	0.010	0.008	0.007	0.006	0.005
1.46	0.009	0.008	0.006	0.005	0.004
1.48	0.009	0.007	0.005	0.004	0.004
1.50	0.008	0.006	0.005	0.004	0.003
1.55	0.006	0.005	0.004	0.003	0.003
1.60	0.005	0.004	0.003	0.002	0.002
1.65	0.004	0.003	0.002	0.002	0.001
1.70	0.003	0.002	0.002	0.001	0.001
1.75	0.002	0.002	0.002	0.001	0.001
1.80	0.002	0.001	0.001	0.001	0.001
1.85	0.002	0.001	0.001	0.001	0.001
1.90	0.001	0.001	0.001	0.001	0.000
1.95	0.001	0.001	0.001	0.000	0.000
2.00	0.001	0.001	0.000	0.000	0.000
2.10	0.001	0.000	0.000	0.000	0.000
2.20	0.000	0.000	0.000	0.000	0.000
2.3	0.000	0.000	0.000	0.000	0.000
2.4	0.000	0.000	0.000	0.000	0.000
2.5	0.000	0.000	0.000	0.000	0.000
2.6	0.000	0.000	0.000	0.000	0.000
2.7	0.000	0.000	0.000	0.000	0.000
2.8	0.000	0.000	0.000	0.000	0.000
2.9	0.000	0.000	0.000	0.000	0.000
3.0	0.000	0.000	0.000	0.000	0.000
3.5	0.000	0.000	0.000	0.000	0.000
4.0	0.000	0.000	0.000	0.000	0.000

Table 8-4 *Cont'd*

u \ N	8.2	8.6	9.0	9.4	9.8
4.5	0.000	0.000	0.000	0.000	0.000
5.0	0.000	0.000	0.000	0.000	0.000
6.0	0.000	0.000	0.000	0.000	0.000
7.0	0.000	0.000	0.000	0.000	0.000
8.0	0.000	0.000	0.000	0.000	0.000
9.0	0.000	0.000	0.000	0.000	0.000
10.0	0.000	0.000	0.000	0.000	0.000
20.0	0.000	0.000	0.000	0.000	0.000

EXAMPLE 8.19

A 2000 ft long rectangular channel of 18 ft width connects two reservoirs (Figure E8.19). The slope of the channel is 0.0015, its roughness factor is 0.02. The water level in the upper reservoir is constant at 5 ft above the channel bottom at the entrance. The water level in the lower reservoir varies. Develop a discharge delivery curve for the channel as a function of the lower reservoir level.

SOLUTION:

Taking the length and the slope of the channel we can calculate the water elevations with respect to a datum plane at the bottom of

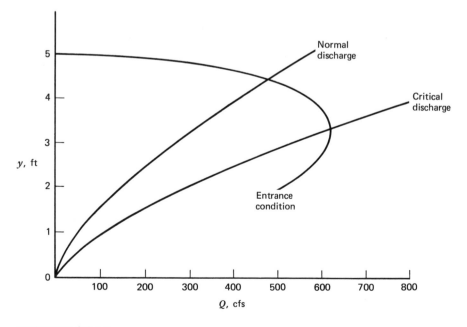

FIGURE E8.19

the exit of the channel denoted as point 2. At point 1, the entrance, the bottom elevation of the channel is

$$h_1 = L \cdot S = 2000 \times 0.0015 = 3 \text{ ft}$$

The upper reservoir elevation is $h_1 + 5 = 3 + 5 = 8$ ft above datum. This corresponds to the zero discharge condition at the lower reservoir level. The maximum discharge in the channel occurs when the depth at the lower reservoir corresponds to critical conditions.

Normal flow in the channel delivers somewhat less. Normal flow occurs when $y_2 = y_1$. The depth y_2 is initially unknown, since h_2 is the total energy level. For lower reservoir elevations above the normal depth y_1 the discharge gradually decreases until it reaches zero when the lower reservoir elevation equals that of the upper reservoir. Using Equation 8.4 we may compute the entrance conditions as follows.

At the upper reservoir with $E_1 = 5$ ft,

$$E_1 = \frac{Q^2}{2gB^2y_1^2} + y_1$$

$$5 = \frac{Q^2}{(64.4)18^2y_1^2} + y_1$$

$$5 = \frac{Q^2}{20,865y_1^2} + y_1$$

This expression showing the entrance conditions may be computed for several depth values and plotted for future reference:

y_1	3	4	4.5	5	ft
Q	612.8	577.8	459.6	0	cfs

For normal flow in the channel, we should solve Equation 8.22 such that the resulting Q_n and y_n correspond to the entrance requirements just stated. For this solution, again, we must compute Q_n for several y_n values, plot the results, and find the intercept of the two graphs.

$$Q_n = \frac{1.49}{n} AR^{2/3}S^{1/2}$$

$$= \frac{1.49}{0.02}(18 \cdot y)\left(\frac{18y}{18+2y}\right)^{2/3} 0.0015^{1/2}$$

$$Q_n = 356.72 \frac{y^{5/3}}{(18+2y)^{2/3}}$$

y	2	3	3.5	4	4.5	ft
Q_n	144	267	336	409	486	cfs

Even without plotting the two curves we note that the normal depth that corresponds to the entrance condition is about

$$y_n = 4.45 \text{ ft}$$

Therefore, the normal discharge that satisfies the entrance condition is

$$Q_n = 475 \text{ cfs}$$

Since the kinetic energy dissipates at the exit the downstream pool elevation at this normal discharge is 4 ft above datum level.

For critical conditions at the exit point Equation 8.6a must be satisfied. This result must also satisfy the entrance conditions. Writing

$$Q_c = b \sqrt{gy^3}$$
$$Q_c = 18(32.2^{1/2})y^{1.5} = 102.14 y^{1.5}$$

and solving it for a series of y values, we have

y	2	3	3.25	3.5	4	ft
Q_c	288	530	597	667	816	cfs

Comparing these results to the entrance conditions we see that for the critical discharge

$$Q_c = 597 \text{ cfs}$$

is a reasonable approximation, if we neglect the acceleration over the channel and the resulting deviation between entrance and exit elevations. In this case the downstream reservoir elevation for maximum discharge is

$$y_c = 3.25 \text{ ft}$$

Now we have three discharge versus downstream elevation points for our delivery curve. For the sake of demonstration we now may compute a fourth one with the aid of the backwater equation.

Let us first assume a discharge such that it is about 50% of the previously computed maximum discharge Q_c, say,

$$Q = 300 \text{ cfs}$$

Now we take a trial value for the depth at the lower reservoir as

$$y_2 = 6 \text{ ft}$$

By inspecting the conditions at the entrance the depth there is about

$$y_1 = 4.75 \text{ ft}$$

Before carrying out the backwater computations, we need the hydraulic exponents. These will be determined for the average depth

$$y_{ave} = \frac{4.75 + 6}{2} = 5.37 \text{ ft}$$

from which $y/b = 5.37/18 = 0.298$ and from Figures 8.5 and 8.10,

$$M = 3 \quad \text{and} \quad N = 2.8$$

From these values, by Equation 8.31,

$$J = \frac{N}{N - M + 1} = 3.5$$

The normal and critical depths associated with the discharge of 300 cfs may be taken off the graph.

The results are

$$y_c = 2.05 \text{ ft}$$

$$y_n = 3.24 \text{ ft}$$

We are now ready to solve Equation 8.35 for B

$$B = \left(\frac{y_c}{y_n}\right)^M \cdot \frac{J}{N} = \left(\frac{2.05}{3.24}\right)^3 \cdot \frac{3.5}{2.8} = 0.32$$

Next we calculate the u and v values by Equations 8.32 and 8.33:

$$u_2 = \frac{6}{3.24} = 1.85$$

$$u_1 = \frac{4.75}{3.24} = 1.47$$

For v's we have $N/J = \dfrac{2.8}{3.5} = 0.8$

$$v_1 = (1.85)^{0.8} = 1.64$$

$$v_2 = (1.47)^{0.8} = 1.36$$

From Table 8-4 now, we may find the corresponding F values

$$F(u_2, 2.8) = 0.198$$

$$F(u_1, 2.8) = 0.325$$

$$F(v_2, 2.8) = 0.252$$

$$F(v_1, 2.8) = 0.393$$

With these values, we enter Equations 8.34 for L_I

$$L_I = \frac{3.24}{0.0015}[(1.85 - 1.47) - (0.198 - 0.325) + 0.32(0.252 - 0.393)]$$

$$= 2160[0.38 + 0.127 - 0.045]$$

$$L_I = 998 \text{ ft}$$

is less than the actual channel length of 2000 ft.

For our second trial we will attempt to increase the resulting L_{II} beyond 2000 ft. For this we will maintain $Q = 300$ but increase y at the downstream end to 7 ft.

Repeating the computations we get a new average depth of

$$y_{ave} = \frac{4.75 + 7}{2} = 5.87 \text{ ft}$$

$$\frac{y}{b} = \frac{5.87}{18} = 0.326$$

$$M = 3 \quad \text{and} \quad N = 2.8.$$

Hence J and B will remain essentially the same.

The new u_2 and v_2 values are

$$u_2 = \frac{7}{3.24} = 2.16 \qquad v_2 = 1.8$$

and

$$F(u_2, 2.8) = 0.146 \qquad F(v_2, 2.8) = 0.198$$

Hence

$$L_{II} = 2160[(2.16 - 1.47) - (0.146 - 0.325) + 0.32(0.198 - 0.393)]$$
$$= 1742 \text{ ft}$$

well approaching $L = 2000$. To solve the problem we now use Equation 8.37.

$$y_0 = \left[\frac{7-6}{1742-998} (2000 - 998) \right] + 6$$
$$= 7.35 \text{ ft}$$

which is the lower reservoir elevation for $Q = 300$ cfs discharge.

Summarizing our results for the downstream reservoir elevations, we have

Q_{cfs}	y_2 ft
0	8.0
300	7.35
475	4.45
620	3.25

These results may now be plotted in the form of a rating curve for the channel with the assumption that the upper reservoir level remains constant. If the level varies, the computations may be carried out for any number of other upstream reservoir elevations.

PROBLEMS

8.1   A rectangular open channel is 20 ft wide and 2 ft deep. The average velocity measured in the channel is 0.5 ft/sec. Calculate the total available energy of the flow.

8.2   Compute the critical discharge for Problem 8.1.

8.3   Determine the critical depth in a rectangular cross section of 12 ft width if the critical discharge is 200 cfs.

8.4   A trapezoidal channel has 6 ft bottom width and $z = 3$ side slopes. Determine the critical depth if the total available energy is 4 ft.

8.5   Determine the critical discharge in a 90 degree triangular cross section if the critical depth is 3 ft.

8.6   In front of a hydraulic jump in a rectangular channel the velocity of the water is 8 m/s and the depth of the water is 40 cm. What is the conjugate depth at the downstream side?

8.7   For the case in Problem 8.6 compute the discharge if the channel is 6 m wide.

8.8   By using Figure 8.8, design an earth channel on a 10^{-4} slope to carry 1.5 m³/s under optimum conditions with side slopes of 1.5 to 1.

8.9   A rock lined channel on 0.0005 slope is rectangular in shape. The bottom width is 8 ft and the depth is 4 ft. Compute the discharge assuming normal flow.

8.10 Determine the conveyance of the channel in Problem 8.9.

8.11 Determine the permissible maximum velocity in a channel in which normal flow takes place at a depth of 4 ft if the soil in which the channel is built is clay.

8.12 The flow in the channel described in Problem 8.9 is dammed up to a level of $y_0 = 18$ ft. Determine the extent to which the backwater effect will be noticeable within 6 in.

8.13 The channel described in Problem 8.9 carries water to a reservoir in which the water level fluctuates between zero and 18 ft. The length of the channel is 4500 ft. At the upstream end the water level is controlled at a constant 6 ft. Develop a discharge curve for the variable downstream depth.

9

FLOW THROUGH HYDRAULIC STRUCTURES

A large share of the hydraulician's work is spent on designing and analyzing hydraulic structures controlling the flow of water in rivers and manmade channels or conducting water through embankments. This chapter surveys the hydraulic design principles and methods related to the most common structures.

Hydraulic structures are built to control or transmit discharges and to maintain water levels in streams and channels. In contrast to previous chapters that deal with the quantitative and qualitative analysis of the flow of water along an essentially uniform channel cross section, this chapter treats localized effects concentrated on or around manmade structures. The types of these structures are many, yet they can be broadly classified into two groups. In one group the flow takes place under pressure through a definitely fixed cross section, in a manner somewhat analogous to pipe flow. Flow through orifices, nozzles, short pipes, sluiceways, or under gates are of this kind. The other group contains cases in which the flow occurs through an initially undetermined cross section as in open channels. These include flow over weirs, spillways, chutes, and drop structures.

In both cases the flow is primarily controlled by the upstream water level. However, if the energy grade line at the downstream side is relatively high, it also exerts influence on the flow through the structure. The difference in total available energy level before and after the structure represents an energy loss. The lost hydraulic energy is dissipated on or immediately behind the structure, as shown in Figure 9.1, in the form of viscous shear within the water particles or within the water and the surrounding structural elements. This effect should be seriously considered in the design as it exerts forces on the structure or it may cause possible erosion. Erosion and underestimation of maximum encountered discharge are the most common cause of failure of hydraulic structures.

The analysis of flow through hydraulic structures is based on the energy equa-

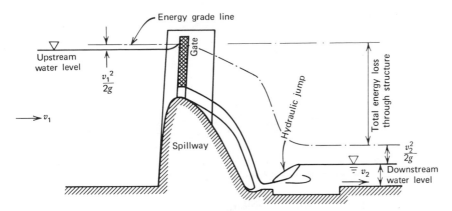

FIGURE 9.1 Energy parameters in flow through a hydraulic structure.

tion. Localized effects of inertia and viscous shear are commonly included in the form of experimentally determined flow coefficients. For simple structures many of these are readily available in hydraulic literature. A representative collection of these data will be included in the following pages. For more complicated structures usually no reliable data is available. For structures of particularly large size and cost the designers often rely on laboratory studies on reduced scale models. To design and operate such small scale models, and to translate their data to full prototype scale the knowledge of modeling laws is essential.

ORIFICES AND SLUICEWAYS

In case a barrier is placed in a stream in which the flow takes place through a geometrically fixed opening located under the upstream water level the flow is analyzed by the orifice formula. Consider a small opening called *orifice* in the side or the bottom of a container as shown in Figure 9.2. The total available energy at the center of this opening equals the depth of the water h. Under the influence of this energy a jet of water will issue out of the orifice with a velocity v. Writing the energy equation, Equation 2.9 for two

points, one at the water surface of the container at point 1, another at the center of the jet at point 2, we obtain

$$\frac{v_1^2}{2g} + \frac{p_1}{\gamma} + Z_1 = \frac{v_2^2}{2g} + \frac{p_2}{\gamma} + Z_2$$

where the following assumptions may be made: Because of the large size of the container the approach velocity at point 1 is negligible when compared to the velocity of the jet at point 2: hence $v_1 \approx 0$. The pressure at point 1 as well as in the thin jet at point 2 equals the atmospheric pressure, hence, $p_1 = p_2$. Substituting $Z_1 - Z_2 = h$, the depth of the orifice below the surface one gets

$$h = \frac{v_2^2}{2g}$$

or

$$v_{\text{jet}} = \sqrt{2gh} \qquad (9.1)$$

which is referred to as Toricelli's equation. The discharge through an orifice can be calculated if the area of the orifice A is sufficiently small with respect to the size of the container, in which case the variation of h from the bottom to the top of the opening is negligible. In this case the velocity of the flow throughout A can be considered constant and the discharge, at least theoretically, is

$$Q = Av = A\sqrt{2gh}$$

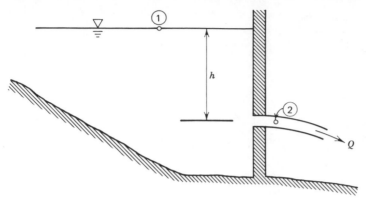

FIGURE 9.2   Notations for the orifice problem.

Experimental results showed that the actual discharge through an orifice is somewhat less than the discharge that would be obtained theoretically. There are two reasons for this discrepancy. First, the size of the jet is somewhat smaller than the size of the opening because of the effect of the curvature of the approaching stream lines in the container, as depicted in Figure 9.3. The narrowest portion of the jet stream is called the "vena contracta." Another cause for the reduction of discharge is the viscous shear effect between the edge of the orifice and the water. This causes a loss of velocity around the edges, as shown in Figure 9.4, with the corresponding decrease of discharge. The com-

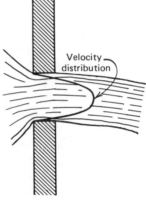

FIGURE 9.4   Velocity distribution due to viscous shear loss at the orifice edge.

bined effect of these two factors gives rise to a *discharge coefficient*, *c*, whose values, based on many experiments with small, round, and square orifices discharging into air, range from 0.6 to 0.68. The experimental results indicate that the discharge coefficient is larger with smaller diameters and larger heads. Including the discharge coefficient *c* into the theoretical equation shown above, one obtains the discharge formula for flow through openings under pressure in the form of

$$Q = cA\sqrt{2gh} \qquad (9.2)$$

In the case when the jet issuing through the opening exits into the downstream water, the value of *h* in the above equation is the difference between the two water levels. For actual structures where

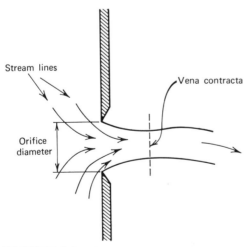

FIGURE 9.3   The development of the vena contracta.

the orifice cannot be considered infinitely thin but, rather, is a short tube, the value of the discharge coefficient increases. Figure 9.5 shows discharge coefficients based on experimental data on short tubes of various materials and configurations.

A practical application of this concept is for sluiceways under dams. A *sluiceway* is a pipe or tunnel that passes through a dam allowing the removal of water from the reservoir as needed. The location of the sluiceway is dependent on the size and type of the dam. Concrete or masonry dams may be equipped with one or more sluiceways at any location determined by reservoir control considerations. In earthdams or rockfill dams the sluiceway must be located outside the limits of the embankment. Seepage along the outside of the pipe walls must be minimized by placing projecting collars on the pipe in certain intervals along its length. The size and number of these collars should increase the seepage path by at least 25%.

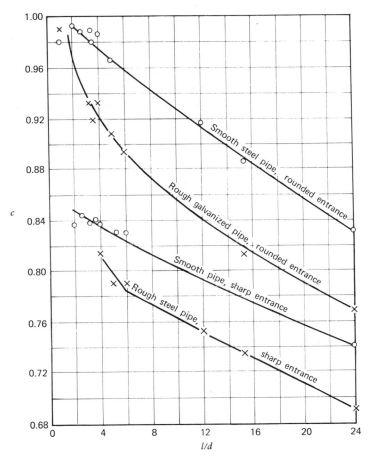

FIGURE 9.5 Discharge coefficients for short tubes.

EXAMPLE 9.1

Determine the required diameter of a 3 ft long horizontal drain pipe located at the bottom of a concrete storage tank, 15 ft below the water level, under the condition that the discharge through the pipe be 9000 gpm.

SOLUTION:

By Equation 9.2 the pipe diameter may be computed as

$$D = \left(\frac{4Q}{c\pi\sqrt{2gh}}\right)^{1/2}$$

where

$$Q = \frac{9000}{449} = 20 \text{ cfs}$$

$$h = 15 \text{ ft}$$

The value of c may be assumed to be 0.6. Therefore,

$$D = \left(\frac{4(20)}{0.6\pi\sqrt{64.4(15)}}\right)^{1/2} = \left(\frac{80}{58.58}\right)^{1/2} = \sqrt{1.36} = 1.16 \text{ ft}$$

$$D = 1.16(12) = 14 \text{ in.}$$

FLOW UNDER GATES

Flow under a *vertical gate* can be defined as a square orifice problem as long as the opening height, a, under the gate is small when compared to the upstream energy level H_0, and the downstream water level H_2 does not influence the flow. By Equation 9.2 we may write

$$Q = bac\sqrt{2g(H_0 - H_1)} \qquad (9.3)$$

in which b is the width of the gate and the other terms are as shown in Figure 9.6. Direct use of this equation is difficult in practice because of the uncertainty in the determination of H_1, the depth of water at the vena contracta. Since H_1 depends on the opening height a, we may write

$$H_1 = \psi a$$

Experimental values for ψ were found to depend on H_0/a and are shown in Figure 9.7.

To eliminate the difficulty in the use of Equation 9.3, Francke rewrote the dis-

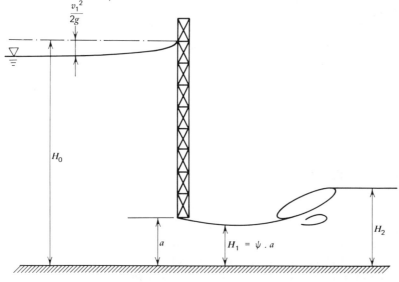

FIGURE 9.6 Notations for flow under gates.

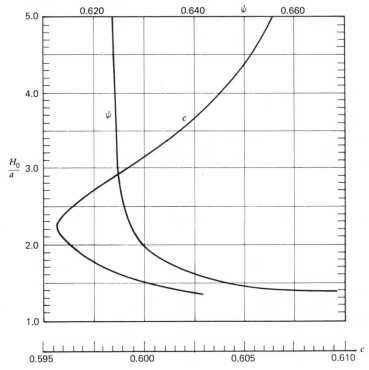

FIGURE 9.7   Discharge coefficient for flow under gates.

charge formula in the form of

$$Q = bac\sqrt{2g}\,\frac{H_0}{\sqrt{H_0 + \psi \cdot a}} \quad (9.4)$$

Experimentally determined values for the discharge coefficient c and for the factor ψ are shown in Figure 9.7. For H_0/a values exceeding the range of the graph, the c value approaches 0.7, and ψ approaches 0.624.

EXAMPLE 9.2

A 6 ft wide vertical gate on the top of a spillway withholds a 4 ft deep water. Determine the discharge under the gate if it is raised by 1 ft.

SOLUTION:

Using Equation 9.4 with

$$b = 6.0 \text{ ft}$$
$$a = 1.0 \text{ ft}$$
$$H_0 = 4.0 \text{ ft}$$

we have

$$Q = 6(1.0)c\sqrt{64.4}\left(\frac{4}{\sqrt{4 + \psi}}\right)$$

The coefficients c and ψ may be obtained from Figure 9.7 with

$$\frac{H_0}{a} = \frac{4}{1} = 4$$

obtaining

$$\psi = 0.624$$

$$c = 0.6036$$

Hence

$$Q = 6\,(0.6036)\,8.025\,\frac{4}{\sqrt{4.624}} = 54.06 \text{ cfs}$$

EXAMPLE 9.3

Determine the required opening of the gate described in Example 9.2 for a discharge of 20 cfs.

SOLUTION:

Entering Equation 9.4 with our known values, we have

$$20 = 6ac\,\sqrt{64.4}\,\frac{4}{\sqrt{4 + \psi a}}$$

Simplifying we have

$$\frac{20}{24\sqrt{64.4}} = 0.103 = \frac{ac}{\sqrt{4 + \psi a}}$$

By observing Figure 9.7 we note that a trial and error solution is possible by assuming the value of a.

Assuming $a = 0.5$, $H_0/a = 8$, and $\psi = 0.6245$, $c \approx 0.665$ approximately. Hence

$$\frac{0.5(0.665)}{\sqrt{4 + 0.5(0.6245)}} = 0.16 > 0.103$$

Therefore a must be further reduced. For our second trial let's take $a = 0.4$ for this:

$$\frac{H_0}{a} = \frac{4}{0.4} = 10.0$$

and the values of the coefficients may be assumed to be $\psi = 0.6245$ as before and $c = 0.7$. Therefore

$$\frac{0.4(0.7)}{\sqrt{4 + 0.4(0.6245)}} = 0.136 > 0.103$$

which is still too large. Next, assume $a = 0.3$. $H_0/a = 13.3$ a very large number for which the coefficients are as before. Hence

$$\frac{0.3(0.7)}{\sqrt{4 + 0.3(0.6245)}} = 0.1026$$

which nearly equals 0.103. Hence our assumption for $a = 0.3$ ft was correct. At this opening the gate delivers 20 cfs discharge.

Equation 9.4 is valid only if the downstream water level is not influencing the flow. This is true when a hydraulic jump exists below the gate. This requires that the conjugate depth of H_2 downstream water level be equal or larger than H_1. The concept of conjugate depth is introduced in Chapter 8. It is calculated by using

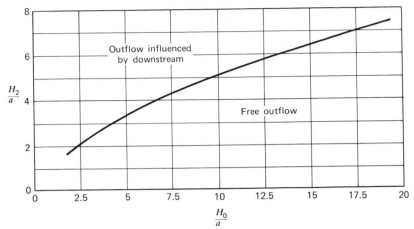

FIGURE 9.8   The range of downstream influence on flow under gates.

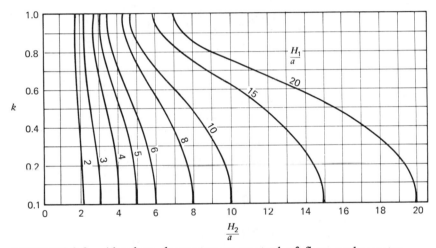

FIGURE 9.9   Absolute downstream control of flow under gates.

Equation 8.8 with $y_2 = H_2$; y_1 is the depth immediately in front of the jump. Instead of carrying out the computations one may utilize Figure 9.8, which gives the relationship between H_0, H_2, and a at the condition when the free outflow becomes retarded by the downstream water level. The discharge under a gate in the case of such *partial downstream control* is computed from

$$Q_{\text{retarded}} = kQ \qquad (9.5)$$

where Q is to be obtained from Equation

9.4, and k is a coefficient that may be found in Figure 9.9 for various H_0, H_2, and a values.

When $H_0 - H_1$ equal or less than H_2 one speaks of *absolute downstream control*. In this case H_1 is completely submerged and the discharge computation is based on $h = H_0 - H_2$. In this case the discharge formula is

$$Q = bac\sqrt{2gh} \qquad (9.6)$$

For metric values this formula may be solved by the graph shown in Figure 9.10.

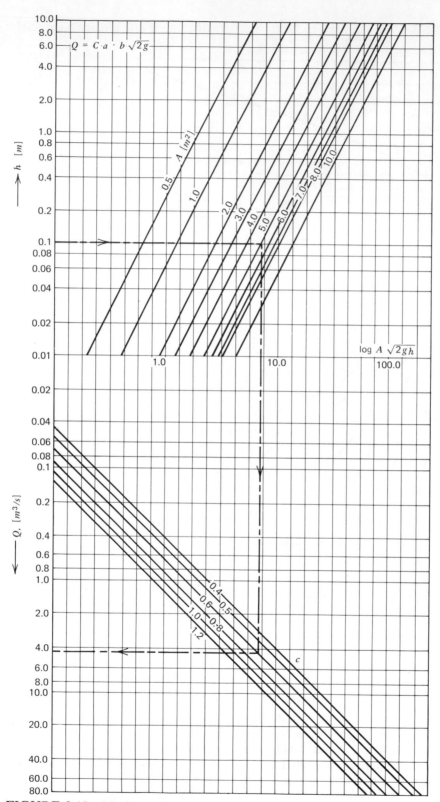

FIGURE 9.10   Nomograph for the solution of the discharge formula for flow under
gates, Equation 9.6.

EXAMPLE 9.4

A 10 ft wide vertical gate discharges into a pool in which the water level is 6 ft. The upstream water level is 8 ft and the gate opening is 2.0 ft. Determine the discharge through the structure.

SOLUTION:

By Figure 9.8 we find the type of flow condition existing.

$$\frac{H_2}{a} = \frac{6}{2.0} = 3.0 \qquad \frac{H_0}{a} = \frac{8}{2} = 4.0$$

Observing the location of the point described, we note that the outflow is influenced by the downstream.

For partially retarded discharge we have to use Equation 9.5. Q is to be determined by Equation 9.4 and corrected by a coefficient k from Figure 9.9. The correction factor is $k = 0.6$.

$$Q = 10(2)c\sqrt{64.4}\left(\frac{8}{\sqrt{8 + \psi 2}}\right)$$

and, from Figure 9.7,

$$\psi = 0.624 \qquad \text{and} \qquad c = 0.6036$$

Therefore the free discharge is

$$Q = 1284 \cdot \frac{0.6036}{\sqrt{8 + 2(0.624)}} = 254.85 \text{ cfs}$$

The retarded discharge is then

$$Q_r = kQ = 0.6(254.85) = 152.9 \text{ cfs}$$

EXAMPLE 9.5

A gate 4 m wide is closing an irrigation canal that is 3 m deep. Determine the head loss through the gate if the gate is opened to a height of 20 cm and the discharge is 1 m³/s. Assume a discharge coefficient of 0.7.

SOLUTION:

Using Equation 9.6 in its graphical form shown in Figure 9.10, we enter at 1 m³/s on the coordinate line showing the discharges, and turn upward at $c = 0.7$. The area of the opening is $b = 10$ m times $a = 0.2$ m, which is 2 m². Turning toward the left at the 2 m² line, we find that $h = 0.02$ m or 2 cm is the head loss through the gate.

WEIRS

Flow taking place over a hydraulic structure under free surface conditions is analyzed with the *weir formula*. Generally speaking all barriers on the bottom of the channel that cause the flow to accelerate in order to pass through can be considered as weirs. More specifically weirs are constructed with openings of simple geometrical shapes. Rectangular, triangular, or trapezoidal shapes are the most common. In each case the bottom edge of the opening over which the water flows is called the *crest*, and its height over the bottom of the reservoir or channel is known as the crest height. The French term *nappe* (sheet) is often applied to the overfalling stream of water. Weirs where the downstream water level is below the crest allow the water to pass in a free fall. Under this condition weirs are good flow measuring devices, particularly if their wall at the crest and sides is thin. Weir shapes other than rectangular are almost exclusively used in flow measurement. The use of *sharp crested weirs* in flow measurement will be discussed in the following chapter. For practical purposes as long as their crest thickness is more than 6/10 of the nappe thickness weirs should be considered broad-crested. Flow over *broad crested weirs* is significantly influenced by viscous drag, which causes a boundary layer to form in the velocity profile of the overflow. This effect is enumerated in the form of a discharge coefficient depending on the shape of the crest, and on the upstream energy level. If the width of the upstream channel is greater than the width of the weir opening we speak of *contracted weirs*. The side contraction in weirs results in an additional contribution to the discharge coefficient by reducing the flow further. In weirs whose width equals the width of the upstream channel surface the side contraction is suppressed. These weirs are called *suppressed weirs*. For a rectangular weir with freely falling water the velocity at any point over the crest can be determined by the energy equation, Equation 2.9. The variables of the problem are also shown in Figure 9.11. The total available energy on the upstream side is

$$E_{\text{upstream}} = h_1 + \frac{v_1^2}{2g} \qquad (9.7)$$

where h_1 is the depth of water on the upstream side over the crest, back at a point where the water level is unaffected by the surface curve, and v_1 is the approach velocity in the channel. When the P height of the crest is significant, as in most hydraulic structures, the approach velocity is relatively small; hence the kinetic energy on the upstream side may be neglected. For free fall the discharge Q will occur under critical flow condition. By virtue of Equation 8.5a the depth of the nappe will be the critical depth, that is,

$$y_2 - y_1 = \tfrac{2}{3}[E_{\text{upstream}}]$$

or, when neglecting the approach velocity and considering $P = y_1$ as our datum plane,

$$y_2 - y_1 = \tfrac{2}{3}h_1 \qquad (9.8)$$

The velocity at any point between y_2 and y_1 will be determined by the energy available at that point. For example, in a thin horizontal strip of just below y_2 the velocity equals

$$v = \sqrt{2g(h_1 - y_2)}$$

For the area of the thin strip of width Δy, shown in Figure 9.11, the elementary discharge is

$$\Delta q = b(\Delta y)v \cong b(\Delta y)\sqrt{2g(h_1 - y_2)}$$

Summing the whole area of the nappe composed of many thin strips between y_1 and y_2 can be performed by a simple step of integral calculus. The result is the common form of the weir formula

$$Q = cb\sqrt{2g}h_1^{3/2} \qquad (9.9)$$

in which Q is the total discharge over the rectangular weir of b width under a head

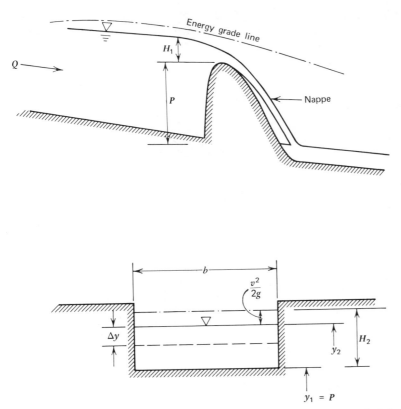

FIGURE 9.11   Notations for the weir formula. P = crest height and b = weir length.

of h_1. The parameter c in this formula is the discharge coefficient of the weir. Equation 9.9 may be solved in metric terms by the aid of the graph shown in Figure 9.12. The formula plotted in Figure 9.12 combines the value of c with the metric value of $\sqrt{2g}$ in the form of $Q = MbH^{3/2}$ where

$$M = c\sqrt{2g} = \frac{Q}{bH^{3/2}} \qquad (9.10)$$

which may be called the *metric weir coefficient*. Table 9-1 shows an extensive collection of various broad crested weir shapes with their metric weir coefficients determined by laboratory experiments. To obtain the c coefficient for computations with English units, we divide the M values found in the table by 4.43. When the downstream water level h_2 exceeds the crest height, it may influence the discharge over the weir. In this case the

water is prevented from passing by free fall. There are two such conditions. As long as the downstream water level is below the midpoint of the nappe over the crest, a hydraulic jump will form over a weir (see Figure 9.13a). The nappe in this case will consist of a supercritical portion preceding the hydraulic jump. With further increase of the downstream water level the jump will first become submerged. Additional increase of h_2, the downstream water level will cause a rise in the upstream water level; the water will be backed up. For this case shown in Figure 9.13b, the discharge could still be computed by Equation 9.9 with the stipulation that h_1 in the equation be replaced by the difference in the measured upstream and downstream water levels. In all cases when the free fall over the weir is prevented by the height of the downstream water level, the discharge coefficient will be reduced. The actual ranges of various flow patterns over weirs depend on the configuration of the weir. For one typical weir shape the ranges of free over-fall, free or submerged hydraulic jump, or subcritical overflow are shown in Figure 9.14.

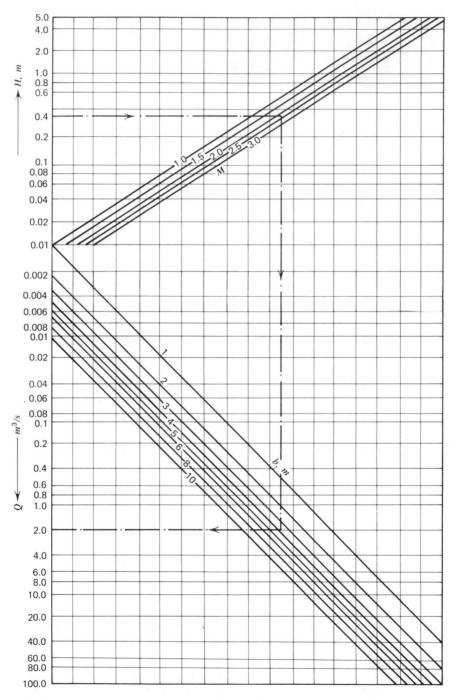

FIGURE 9.12 Nomograph for the solution of the weir discharge formula, Equation 9.10. $Q = MbH^{3/2}$.

Table 9-1 Discharge Coefficients for Broad Crested Weirs[a]

	Cross section	0.15	0.30	0.45	0.60	0.75	0.90	1.20	1.50
		Upstream head h [m]							
1	0.8; 5:1; 0.2; 2:1	1.61	1.86	1.98					
2	0.8; 5:1; 0.2; 3:1	1.60	1.80	1.90					
3	0.8; 5:1; 0.2; 5:1	1.58	1.75	1.79					
4	0.8; 5:1; 0.4; 2:1	1.53	1.64	1.77					
5	0.8; 5:1; 0.4; 6:1	1.54	1.62	1.69					
6	0.8; 2:1; 0.2; 2:1	1.72	1.88	1.98					
7	0.8; 1:1; 0.2; 2:1	1.65	1.88	2.00					
8	0.8; 0.2; 2:1	1.53	1.80	1.93					
9	1.6; 2:1; 0.1; 2:1				1.96	1.96	1.97	1.99	2.02
10	1.6; 2:1; 0.1; 5:1				1.94	1.92	1.89	1.92	1.97
11	2:1; 0.1; 1.6		2.12	2.10	2.08	2.08	2.06	2.04	2.00
12	2:1; 0.2; 1.6		1.88	1.96	2.01	2.04	2.05	2.05	2.05
13	3:1; 0.2; 1.6				1.96	1.96	1.96	1.96	1.96
14	5:1; 0.2; 1.6				1.86	1.86	1.86	1.86	1.86
15	0.8; 1:1; 0.15; 1:1	1.81	2.00						
16	0.8	2.10	2.35						
17	0.22	1.57	1.73	1.80	1.82	1.83	1.83		
18	0.6	1.44	1.46	1.55	1.56	1.69	1.76	1.84	
19	1.5	1.43	1.47	1.45	1.46	1.47	1.46	1.48	1.59
20	4.5	1.48	1.45	1.44	1.44				
21	1.5; 0.1; 0.83		1.56	1.60	1.65	1.70	1.74	1.84	1.92
22	1.5; 0.1; 2.15		1.56	1.56	1.55	1.55	1.55	1.55	1.54
23	0.8; 1:1	2.13	2.13	2.13					
24	0.8; 2:1	1.93	1.94	1.94					
25	0.54; 2:1	1.94	1.98	1.97					

[a] All dimensions are in meters. Tabulated values represent metric weir coefficients.

Table 9-1 *Cont'd*

	Cross section	0.15	0.30	0.45	0.60	0.75	0.90	1.20	1.50
					Upstream head h [m]				
26		1.69	1.73	1.73					
27		2.28	2.25	2.06					
28		2.08	2.12	2.12					
29		1.92	1.93	1.92					
30		2.10	2.13	2.13					
31		2.03	2.03	2.01					
32		2.03	2.03	2.01					
33		1.65	1.94	2.10					
34			1.72	1.76	1.76	1.76	1.76	1.76	1.76
35			1.87	1.84	1.81	1.82	1.82	1.85
36			1.91	1.90	1.87	1.84	1.83	1.86	1.90
37					1.89	1.87	1.87	1.88
38			1.81	1.81	1.82	1.86	1.90	1.97	2.01
39		1.13	1.82	1.83	1.85	1.87	1.88	1.95	2.05
			1.89	1.90	1.88	1.88	1.90		
40		1.76	1.86	1.90	1.93	1.96	1.97	2.03	2.11
			1.92	1.98	2.06	2.04	2.04		
41		1.72	1.90	2.00	2.06	2.10	2.13		
42		1.78	1.84	1.89	1.93	1.97	2.00		
43		1.75	1.81	1.85	1.88	1.90	1.92	1.95	
44		1.80	1.92	1.95	1.94	1.85	1.82		
45		1.94	1.94	1.95	1.92	1.85	1.81	1.79	
46		1.72	1.72	1.70	1.72	1.76	1.79	1.85	
47		1.70	1.71	1.82					
48			2.09						

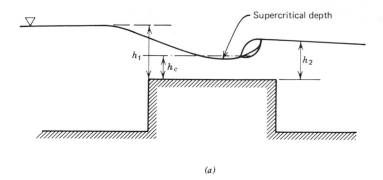

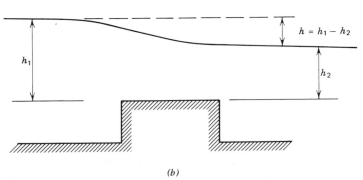

FIGURE 9.13 Noncritical flows over broad crested weirs. (a) Weir with hydraulic jumps and (b) Weir with subcritical flow.

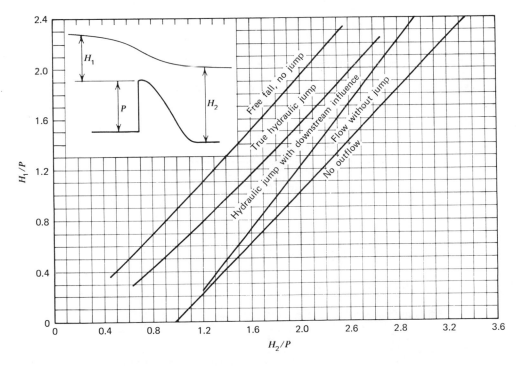

FIGURE 9.14 Ranges of flow types over weirs.

EXAMPLE 9.6

A rectangular broad crested weir is 30 ft long and is known to have a discharge coefficient of 0.7. Determine the discharge if the upstream water level is 2 ft over the crest.

SOLUTION:

Equation 9.9 applies. Substituting the variables we have

$$Q = cb\sqrt{2g}h_1{}^{3/2}$$
$$= 0.7(30)\sqrt{64.4}(2^{3/2})$$
$$= 476 \text{ cfs}$$

EXAMPLE 9.7

A weir 10 m long with a cross-sectional design similar to case number 4 of Table 9-1 is flowing with an upstream head of 45 cm. Determine the discharge.

SOLUTION:

From Table 9-1 the metric discharge coefficient at 45 cm upstream head is $M = 1.77$.

Using Figure 9.12 with $b = 10$ m, $H = 0.45$ m, and $M = 1.77$ the discharge is found to be

$$Q = 4.5\,\frac{\text{m}^3}{\text{s}}$$

EXAMPLE 9.8

A 55 ft weir has a crest configuration similar to case number 45 in Table 9-1. What is the discharge through the weir at a head of one foot of water?

SOLUTION:

To use the metric table, we convert the upstream head into meters, resulting in 30 cm head for which $M = 1.94$.

Converting M into a c coefficient requires that we divide it by 4.43 resulting in

$$c = \frac{1.94}{4.43} = 0.438$$

Using Equation 9.9 we have

$$Q = 0.438(55)\sqrt{64.4}(1)^{3/2} = 193.32 \text{ cfs}$$

EXAMPLE 9.9

A 20 ft long weir was measured independently to carry a 50 cfs discharge when the crest is overtopped by 8 in. of water. Determine the discharge coefficient of the weir.

SOLUTION:

Rearranging Equation 9.9 in the form of

$$c = \frac{Q}{b\sqrt{2gh}^{3/2}}$$

and substituting the known variables, we obtain

$$c = \frac{50}{20\sqrt{64.4}(8/12)^{1.5}}$$
$$= 0.57$$

is the discharge coefficient sought.

EXAMPLE 9.10

A dam's weir is designed to have a crest height of 16 ft. The maximum upstream head over the crest should not exceed 7 ft. However, the length of the weir is to be selected such that the downstream water level does not influence the upstream conditions, and therefore, does not retard the flow during critical floods. What is the allowable maximum downstream water level under these conditions?

SOLUTION:

Figure 9.14 may be used with

$$\frac{H_1}{P} = \frac{7}{16} = 0.437$$

The right bounding line of the region of true hydraulic jump represents the limiting condition for no downstream influence. In our case it is

$$\frac{H_2}{P} = 0.55$$

Therefore, with $P = 16$ the limiting value is

$$H_2 = 8.8 \text{ ft}$$

EXAMPLE 9.11

A weir is to be placed in a channel in order to measure the discharge; we use a rating curve to be developed after construction by independent discharge measurements. The normal flow in the channel under extreme conditions will not exceed 4 m. What should be the minimum crest height of the weir?

SOLUTION:

By using Figure 9.14 we consider the limit of the free fall condition, the uppermost line in the graph. Developing the equation for this line, we determine that its slope is 0.92 and its intercept at

$H_2/P = 0$ is -0.1; therefore,

$$\frac{H_1}{P} = -0.1 + 0.92 \frac{H_2}{P}$$

from where $P = 9.2H_2 - 10H_1$.

Next we establish the normal depth of the flow in the downstream channel under the maximum expected discharge, which gives our maximum value for H_2. In our case H_2 equals 4 m. Therefore, the minimum required crest height is

$$P = 36.8 - 10H_1$$

where H_1 is determined by Equation 9.9 by using the maximum discharge expected. In order to minimize the required P crest height or to keep H_1 within a prescribed limit, the crest width b may be suitably selected in Equation 9.9.

SPILLWAYS

In order to allow the excess water to pass over the dam in a safe manner, dams are equipped with spillways. A *spillway* is a rectangular concrete channel connecting upstream to downstream over which the water flows at a supercritical velocity. For ideal shape, spillways should follow the underside of the overfalling nappe of water. Theoretically it is of parabolic configuration with a reversed curved lower portion, the bucket, that serves to deflect the falling water smoothly toward the downstream side. It is important that for all discharges the descending water be in contact with the spillway surface. Otherwise, if the water is allowed to shoot away from the spillway, it may exhibit hydraulic instability. The resultant cyclic pounding on the structure causes significant destructive effects.

The discharge of spillways may be calculated by the corresponding broad crested weir formula Equation 9.9 with proper selection of the discharge coefficient. For flows less than the design discharge, the discharge coefficient of the spillway will be less. As the nappe falls over the spillway, it is subjected to the gravitational acceleration. As the velocity increases downward the thickness of the nappe correspondingly decreases. The shape of the nappe in free fall may be determined, on a theoretical basis, assuming no friction on the weir. The computation is facilitated by the coordinates developed by Bazin and Craeger on the basis of the dynamics of free fall and shown in Table 9-2. The notations of this table correspond to those shown in Figure 9.15 with $X = xH_d$ and $Y = yH_d$. The y_1 and y_3 relative coordinates correspond to the lower and upper coordinates of the nappe and y_2 refers to the spillway shape recommended by Bazin and Craeger.

Numerous experimental and theoretical studies on spillway nappe profiles resulted in several standard spillway shapes. A typical one, developed by the U.S. Army Engineers' Waterways Experiment Station, is represented by the equation

$$X^n = (KH_d^{n-1})Y \qquad (9.11)$$

where X and Y are the horizontal and vertical coordinates of the spillway, respectively, with their origins at the highest point of the crest, as shown in Figure 9.15, and H_d is the design head upstream, excluding the kinetic energy of the approach velocity. The design head corresponds to the maximum expected discharge. The coefficients K and n depend on the up-

Table 9-2 Craeger Coordinates for Contoured Spillway and Nappe Configuration

Horizontal Coordinate x	Lower Nappe Surface y_1	Spillway Surface y_2	Upper Nappe Surface y_3	Horizontal Coordinate x	Lower Nappe Surface y_1	Spillway Surface y_2	Upper Nappe Surface y_3
0	0.126	0.126	− 0.831	2.1	1.456	1.369	0.834
0.1	0.036	0.036	− 0.803	2.2	1.609	1.508	0.975
0.2	0.007	0.007	− 0.772	2.3	1.769	1.654	1.140
0.3	0.000	0.000	− 0.740	2.4	1.936	1.804	1.310
0.4	0.007	0.006	− 0.702	2.5	2.111	1.960	1.500
0.5	0.027	0.025	− 0.655	2.6	2.293	2.122	1.686
0.6	0.063	0.060	− 0.620	2.7	2.482	2.289	1.880
0.7	0.103	0.098	− 0.560	2.8	2.679	2.463	2.120
0.8	0.153	0.147	− 0.511	2.9	2.883	2.640	3.390
0.9	0.206	0.198	− 0.450	3.0	3.094	2.824	2.500
$X = H_d \rightarrow 1.0$	0.267	0.256	− 0.380	3.1	3.313	3.013	2.70
1.1	0.355	0.322	− 0.290	3.2	3.539	3.207	2.92
1.2	0.410	0.393	− 0.219	3.3	3.772	3.405	3.16
1.3	0.497	0.477	− 0.100	3.4	4.013	3.609	3.40
1.4	0.591	0.565	− 0.030	3.5	4.261	3.818	3.66
1.5	0.693	0.662	+ 0.090	3.6	4.516	4.031	3.88
1.6	0.800	0.764	+ 0.200	3.7	4.779	4.249	4.15
1.7	0.918	0.873	+ 0.305	3.8	5.049	4.471	4.40
1.8	1.041	0.987	+ 0.405	3.9	5.326	4.699	4.65
1.9	1.172	1.108	+ 0.540	4.0	5.610	4.930	5.00
2.0	1.310	1.255	+ 0.693	4.5	17.50	6.460	6.54

To get the X and Y values, multiply tabulated data by H_d.

y_1, y_2, and y_3 are measured vertically from crest.

stream slope of the spillway and are given as follows:

Upstream Slope	K	n
Vertical	2.000	1.850
3 on 1	1.936	1.836
3 on 2	1.939	1.810
3 on 3	1.873	1.776

For small dams, spillways are often made in a straight, flat shape particularly in the case of earth dams when the spillway's concrete mat is laid on the top of the earth fill. The energy gained by the water falling over the spillway could cause significant erosion downstream if not abated. To reduce the excess energy various techniques may be applied. The energy may be reduced at small structures by roughening the spillway and the following apron surface. Concrete blocks and bars and other energy absorbers are used in practice for this purpose. These result in a cascading flow on the spillway that brings about significant energy dissipation.

At the bottom of the dam the spillway is often built with an outward curvature to direct the flow downstream. The curvature of this portion depends on the crest height and on the design head over the weir. Table 9-3 shows the radius of curvature, in meters recommended for various crest heights and heads.

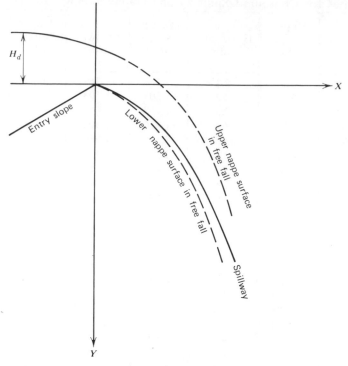

FIGURE 9.15 Interpretation of the notations for flow over spillways (Table 9-2).

Table 9-3 Recommended Radius of Curvatures for Buckets below Spillways[a]

Crest Height P [m]	H [m]						
	0.6	1.5	3.0	4.5	6.0	7.5	9.0
	Radius of Curvature, R, in meters						
6	1.8	3.0	4.5	6.0	7.8	9.0	10.5
9	2.3	3.4	5.1	7.0	8.4	10.0	11.5
12	2.4	4.2	6.0	7.5	9.0	10.7	12.4
15	2.7	4.5	6.6	8.5	9.8	11.4	13.1
30	2.8	6.3	9.6	11.9	13.6	15.0	16.6
45	2.9	7.4	12.0	14.2	16.9	18.7	20.0
60	3.0	7.5	13.5	15.0	20.0	21.6	23.4

[a]From the book of standards of the Research Institute of Water Resources, Budapest, Hungary, written by O. Starosolszky and L. Muszkalay, entitled *Mütárgyhidraulikai Zsebkönyv* (handbook of hydraulic structures, in Hungarian), Vols. 1 and 2, published by the "Technical Bookpublisher Co." Budapest, Hungary, 1961.

214

EXAMPLE 9.12

Design a spillway such that at a maximum flood of $Q = 14{,}000$ cfs the water elevation in the reservoir does not exceed 15 ft over the crest height, which is to be 75 ft. The discharge coefficient is assumed to be 0.62. The entrance slope of the spillway on the upstream side is to be $z = 3$. The curvature of the spillway should conform to the U.S. Army Corps of Engineers formula given in Equation 9.11, connecting to a downstream slope of $z = 5$. Determine the required spillway width and plot the curvature of the spillway. See Figure E9.12.

SOLUTION:

First, calculate the required spillway width for Equation 9.9,

$$Q = cb\sqrt{2g}h^{3/2}$$

with

$$Q = 14{,}000 \text{ cfs}$$
$$c = 0.62$$
$$h = 15 \text{ ft}$$

Hence

$$b = \frac{14{,}000}{(0.62)\sqrt{64.4}(15)^{1.5}} = 48.43 \text{ ft}$$

Select

$$b = 48.5 \text{ ft}$$

For the curvature of the spillway use Equation 9.11 with

$$K = 1.936 \qquad \text{and} \qquad n = 1.836$$

which corresponds to an entry slope of $z = 3$. Then,

$$X^{1.836} = [1.936(15)^{1.836-1}]Y$$

or

$$\frac{X^{1.836}}{9.62} = Y$$

or

$$\left(\frac{X}{15}\right)^{1.836} = 1.936\left(\frac{Y}{15}\right)$$

Computing for a sequence of X values we get

$X =$	0	2	4	6	8	10	12	14	16	18	20
$Y =$	0	0.37	1.33	2.79	4.73	7.13	9.96	13.22	16.89	20.97	25.44

$X =$	22	24	26	28	30	32	34	36	38
$Y =$	30.30	35.55	41.18	47.18	53.55	60.29	67.39	74.85	82.66

These values are plotted in Figure E9.12.

The curvature of the toe of this spillway may be determined from Table 9-3.

Converting h and P into meters (1 ft = 0.3048 m), we get

$$15 \text{ ft} = 4.572 \text{ m}$$
$$75 \text{ ft} = 22.86 \text{ m}$$

From Table 9-3 the radius of curvature is 10.2 m, which is

$$R = 33 \text{ ft}$$

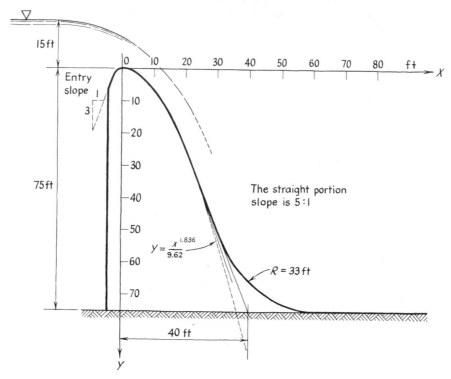

FIGURE E9.12   The straight portion of the slope is 5:1.

EXAMPLE 9.13

A spillway with a 50 ft crest height operates under a 12 ft upstream design head. Using the coordinates recommended by Bazin and Craeger (Table 9-2), determine the location of the outside and inside surfaces of the nappe and the recommended location of the spillway surface at one point, 25 ft below the crest.

SOLUTION:

The coordinates shown in Table 9-2 may be converted to actual measurements by multiplying them with the design head H_d, which in our case is 12 ft. To get the x coordinates needed we find y_1, y_2, and y_3 values such that

$$H_d y = 25 \text{ ft}$$
$$y = \frac{25}{15} = 1.666$$

These values from Table 9-2 correspond to x values as follows:

$$\text{for} \quad y_1 = 1.666 \quad x_1 = 2.22$$
$$\text{for} \quad y_2 = 1.666 \quad x_2 = 2.31$$
$$\text{for} \quad y_3 = 1.666 \quad x_3 = 2.59$$

Multiplying these x coordinates by $H_d = 12$ ft, we obtain

$$X_1 = 2.22(12) = 26.64 \text{ ft}$$

which is the inside surface of the nappe for free fall

$$X_3 = 2.59(12) = 31.08 \text{ ft}$$

which is the outside surface of the nappe for free fall.

Hence the horizontal width of the overfalling sheet of water is

$$X_3 - X_1 = 31.08 - 26.64 = 4.44 \text{ ft}$$

at a location 25 ft below crest.

The location of the spillway surface at this point, recommended by Bazin and Craeger, is

$$X_2 = 2.31(12) = 27.72 \text{ ft}$$

measured horizontally away from the crest and 25 ft vertically below the crest. Since X_1 the inside nappe surface is less than the position of the spillway, it is shown that the spillway is supporting the nappe to some extent.

For a complete spillway design similar values may be computed for other depths below the crest. The position of the outer nappe surface is required in the judicious design of the required side walls bordering the spillway.

ENERGY DISSIPATORS

The velocity of the flow at the toe of the spillway may be computed by

$$v_1 = \sqrt{2g(P + H - y_1)} \qquad (9.12)$$

where P is the crest height, H is the head over the weir including the kinetic energy head, and y_1 is the depth of the nappe at the toe. This velocity is usually of considerable magnitude. Hence it results in a large amount of kinetic energy capable of causing heavy scouring. The situation is identical in case of flow under a gate when the downstream water elevation is low or in the case of a culvert exit. In order to avoid a gradual erosion of the downstream channel and the possible destruction of the structure itself, the excess energy must be dissipated. Assuming that the downstream channel is resistant to scouring and that it is of a near horizontal slope of constant cross section and further providing that no particular effort is made to dissipate the energy of the water leaving the spillway, the following would occur: The flow would shoot out with an initial velocity according to Equation 9.12 in the channel (called apron or tailrace) losing its energy gradually due to channel friction. Since this flow is supercritical, its energy loss rate would be considerable, resulting in a gradual decrease of velocity and a corresponding increase in depth. The flow could be computed with reasonable accuracy by using the Chézy equation although the flow profile is somewhat curved, being a type

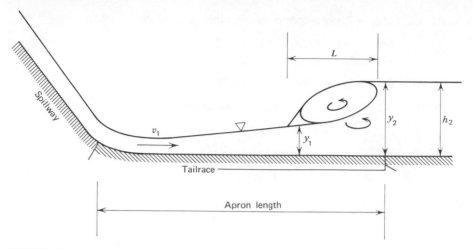

FIGURE 9.16   Notations for a hydraulic jump below a spillway.

of backwater curve. Applying Equation 8.16 and knowing the roughness coefficient, the rate of energy loss may be determined. As the depth of flow increases it will reach a level at which the given discharge Q may flow in the given channel, defined by its slope S and roughness n, either at supercritical or subcritical velocity, by virtue of the concepts depicted in Figure 8.1. The corresponding depths involved here are the conjugate depths described by Equation 8.8. By the law of minimum energy, discussed in Chapter 2, nature will select the flow condition at which the rate of energy loss is less, that is, the flow will change into a subcritical flow. For this reason a hydraulic jump must occur. The energy loss associated with the hydraulic jump is computed from Equation 8.13.

If for a given discharge the downstream channel slope is mild, the downstream water level may be assumed to be of normal depth. This, as shown in Figure 9.16, will define the downstream total energy content of the flow after the jump

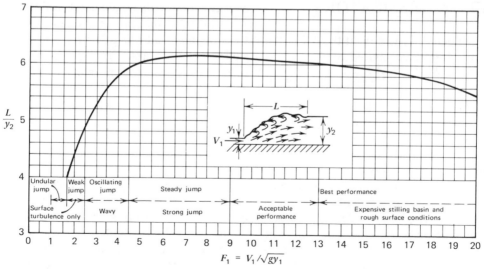

FIGURE 9.17   Ranges of various types of hydraulic jumps. (Courtesy of Bureau of Reclamation.)

and in turn will control the location of the jump. Starting from this control point toward the upstream direction one can make the following conclusions: at the control point the depth of the tail water y_2 equals the subcritical conjugate depth. At the beginning of the jump the depth is then y_1. The length of the jump may be determined by the data plotted in Figure 9.17. The information given in Figure 9.17 is based on experimental data obtained by the U.S. Bureau of Reclamation. With d_1 and Q known, the Froude number at the beginning of the jump may be computed. The remaining uncertainty concerns the distance between the toe of the spillway (or the location of a gate) and the beginning of the jump. The magnitude of this distance is very sensitive to the roughness of the channel. As the construction cost of downstream aprons depend on their required length, the designers should strive to minimize this length. This necessitates that the jump be localized by absorbing the excess energy in the supercritical flow, causing the jump to occur as close to the structure as possible. This end is served by the design and construction of energy dissipators.

EXAMPLE 9.14

A 12 ft wide spillway delivers a discharge of 250 cfs such that the depth of the water at the toe of the spillway is 1 to 2 ft, and the depth of the downstream water is 4 ft (Figure E9.14). Determine the required size of the downstream apron.

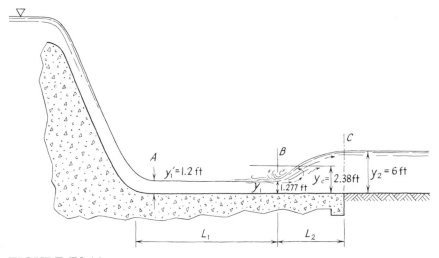

FIGURE E9.14

SOLUTION:

Assume that the spillway is rectangular and identical throughout the apron. From $Q = 250$ cfs, therefore, the critical depth y_c can be calculated. The discharge per unit width $= 250/12 = 20.83$ cfs/ft. Then from

$$y_c = \left[\frac{q^2}{g}\right]^{1/3} = \left[\frac{(20.83)^2}{32.2}\right]^{1/3} = 2.38 \text{ ft}$$

To find the conjugate depth of $y_2 = 6$ ft, we see that

$$\frac{q^2}{g} = \frac{1}{2} y_1 y_2 (y_1 + y_2)$$

$$= \frac{(20.83)^2}{32.2} = \frac{1}{2}(y_1)4(y_1 + 4)$$

$$y_1 = 1.277 \text{ ft}$$

Thus at point B, right in front of the jump the velocity is

$$v_1 = \frac{250}{(12)1.277} = 16.31 \text{ fps}$$

The hydraulic radius

$$R_1 = \frac{12 \times 1.277}{12 + 2 \times 1.277} = 1.053 \text{ ft}$$

At point A, the beginning of the apron $y_1' = 1.2$ ft and

$$v_1' = \frac{250}{12 \times 1.2} = 17.36 \text{ fps}$$

$$R_1' = \frac{12 \times 1.2}{12 + 2 \times 1.2} = 1 \text{ ft}$$

The average velocity between A and B

$$v_{ave} = (17.36 + 16.31)/2 = 16.84 \text{ fps}$$

and

$$R_{ave} = \frac{1 + 1.053}{2} = 1.027 \text{ ft}$$

Now, to find the slope of the energy grade line S slope we assume that the friction coefficient of the apron is $n = 0.014$ and then we write Equation 8.22 as

$$v = \frac{Q}{A} = \frac{1.49 \, R^{2/3}}{n} S^{1/2}$$

$$16.84 = \frac{1.49}{0.014}(1.027)^{2/3} S^{1/2}$$

$$S = \left[\frac{16.84}{(1.027)^{2/3}} \times \frac{0.014}{1.49} \right]^2 = 0.024$$

Hence, the length of the tailrace before the jump is

$$(S_0 - S)L_{AB} = \left(\frac{v_1^2}{2g} + y_1 \right) - \left(\frac{v_1'^2}{2g} + y_1' \right)$$

$$= \left[\frac{(16.31)^2}{2 \times 32.2} + 1.277 \right] - \left[\frac{(17.36)^2}{2 \times 32.2} + 1.2 \right]$$

$$L_{AB} = \frac{5.41 - 5.88}{0 - 0.024} = \frac{-0.47}{-0.024} = 19.58 \text{ ft}$$

The Froude number at B,

$$\mathbf{F}_1 = \frac{V_B}{\sqrt{gy_1}} = \frac{16.31}{\sqrt{32.2 \times 1.277}} = 2.54$$

From Figure 9.17 the length of the jump $L_{BC} = 4.9 \times 4 = 19.6$ ft. Then, the total apron length $= 19.58 + 19.6 = 39.18$ ft.

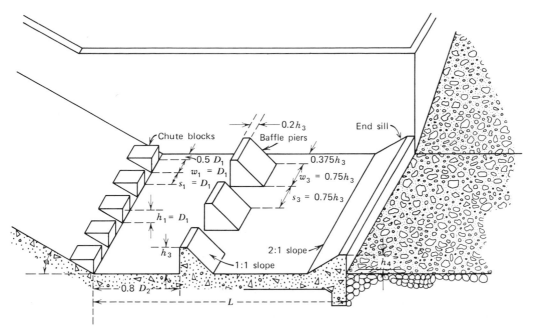

FIGURE 9.18 The U. S. Bureau of Reclamation's Type III energy dissipator. (Courtesy of Bureau of Reclamation.)

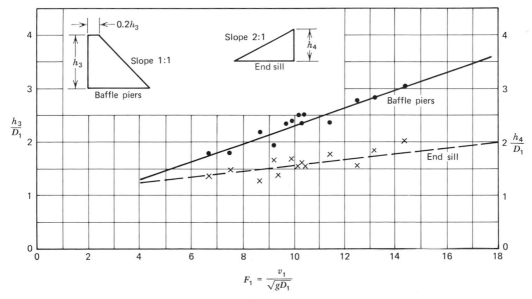

FIGURE 9.19 Sizing chart of energy dissipator components shown in Figure 9.18. (Courtesy of Bureau of Reclamation.)

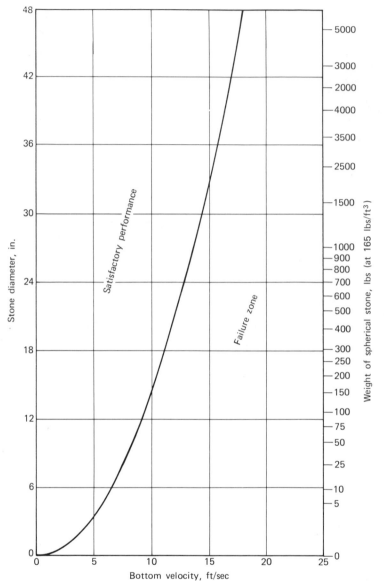

FIGURE 9.20 Recommended size of riprap for erosion control (after Bureau of Reclamation). The riprap should be composed of a well graded mixture, but most of the stones should be of the size indicated by the curve. Riprap should be placed over a filter blanket or bedding of graded gravel in a layer 1.5 times (or more) as thick as the largest stone diameter. Curve shows minimum size stones necessary to resist movement.

A great many different energy dissipators have been built at various types of hydraulic structures. Reports on experimental studies and design recommendations are abundant in the hydraulic literature. Standard designs of energy dissipators used by agencies of the U.S. Federal Government exist. A typical standard design, incorporating several basic energy absorbing features is the U.S. Bureau of Reclamation's Type III basin, which will be reviewed here as an example of an efficient energy dissipator. The design was developed for relatively small structures with a configuration shown in Figure 9.18. The size of the energy dissipator components of this stilling basin is shown in Figure 9.19 as a function of the Froude number and conjugate depth at the beginning (supercritical portion) of the

hydraulic jump. The length of the basin, L, is recommended to be about $2.8d_2$ where d_2 is the subcritical conjugate depth of the jump. This design confines the hydraulic jump within the length of the basin. It also tends to shorten the length of the hydraulic jump by about 50%. It is recommended that the channel below any energy dissipator be protected against erosive currents by crushed stone riprap. The necessary stone size as a function of the expected erosion velocity is presented in Figure 9.20. The data is based on actual field experiences by the U.S. Bureau of Reclamation. Erosion control in a riprap covered channel section following an energy dissipator may be further improved by allowing the channel to flare out to a size several times its normal width.

EXAMPLE 9.15

Determine the required stone size of the riprap to be placed after the apron designed in Example 9.14.

SOLUTION:

Assuming that the velocity immediately below the hydraulic jump is uniform, and computing this value from

$$Q = 250 \text{ cfs}$$
$$\text{Depth} = 4 \text{ ft}$$
$$\text{Width} = 12 \text{ ft}$$
$$v_{\text{ave}} = \frac{250}{4(12)} = 5.21 \frac{\text{ft}}{\text{sec}}$$

Using Figure 9.20 with $V = 5.21$ fps, we find that the required stone diameter is 4 in.

CULVERTS

Highways and railroads traversing the land cut across individual watersheds. To allow the flow from each watershed across the embankment, culverts arc built at the lowest points of the valleys. Although simple in appearance, the hydraulic design of culverts is no easy matter. The hydraulic operation of culverts under the

various possible discharge conditions presents a somewhat complex problem that cannot be classified either as a flow under pressure or as a free surface flow problem. The actual conditions involve both of these basic concepts.

The fundamental objective of hydraulic design of culverts is to determine the most economic diameter at which the design discharge is passed without exceeding the allowable headwater elevation. The major components of a culvert are its inlet, the culvert pipe barrel itself, and its outlet with the exit energy dissipator, if any. Each of these components have a definite discharge delivery capacity. The component having the least discharge delivery capacity will control the hydraulic performance of the whole structure.

One speaks of *inlet control* if, under given circumstances, the discharge of a culvert is dependent only on the headwater above the invert at the entrance, the size of the pipe, and the geometry of the entrance. With the inlet controlling the flow, the slope, length, and roughness of the culvert pipe does not influence the

discharge. In this case, the pipe is always only partly full although the headwater may exceed the top of the pipe entrance and hence the flow enters the pipe under pressure. Figure 9.21 shows a typical nomograph by which the discharge Q could be determined for a culvert of D diameter under a headwater depth HW. The nomograph is for a square-edged entrance in a headwall. Similar nomographs are found in governmental and trade literature for many other entrance conditions. Short culverts with relatively negligible tailwater elevations almost always operate under inlet control. *Outlet control* occurs when the discharge is dependent on all hydraulic variables of the structure.

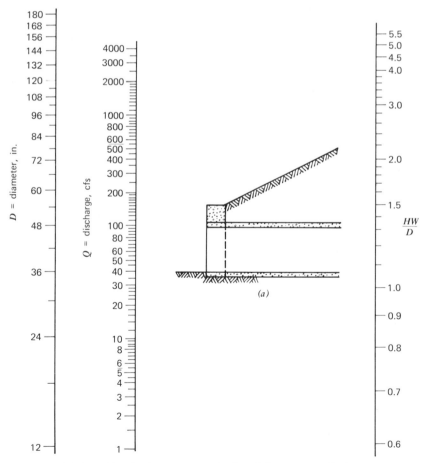

FIGURE 9.21 Typical nomograph for inlet controlled culvert design. (*a*) Square-edged entrance. (From *Handbook of Concrete Culvert Pipe Hydraulics*, Portland Cement Association, 1964.)

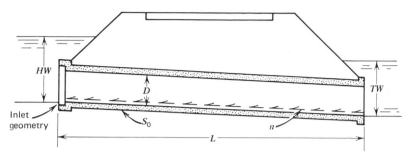

FIGURE 9.22 Notations for culvert analysis.

Figure 9.22 shows the notations relative to these variables. These include the slope S, length L, diameter D, roughness n, tailwater depth TW, headwater depth HW. Unless the tailwater level is above the top of the culvert exit the pipe will be only partly full. This means that the flow in the pipe will be open channel flow.

Flow in partially full circular pipe may be analyzed in the same manner as described in the previous chapter. The depth of the flow d relative to the pipe diameter D determines the area of the flow as well as its hydraulic radius. Figure 9.23 aids in the computation of the various hydraulic parameters in relation to those at full flow. From these data the critical depth d_c may be obtained for any discharge. The analytic complexity of this computation is resolved by the graph presented in Figure 9.23.

Due to the "vena contracta" at the inlet

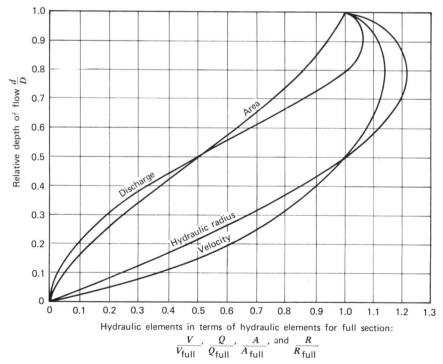

FIGURE 9.23 Hydraulic parameters of circular pipes flowing partially full. (From *Handbook of Concrete Culvert Pipe Hydraulics*, Portland Cement Association, 1964.)

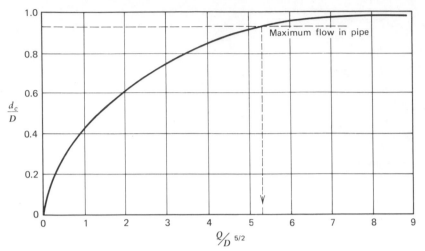

FIGURE 9.24   Critical depth graph for circular pipes flowing partially full. (From *Handbook of Concrete Culvert Pipe Hydraulics*, Portland Cement Association, 1964.)

created by the entrance conditions under the headwater HW, the flow in most cases starts in a supercritical condition in the culvert. The friction on the pipe wall gradually dissipates the energy inherent in the flowing water causing the depth of the flow to increase. Depending on the tailwater elevation the supercritical flow may convert into subcritical flow through a hydraulic jump. As the critical depth is characterized by the fact that it allows the maximum possible discharge to pass under the prevailing total available energy, it is the essence of desirable culvert operation that the flow be under critical conditions. The total available energy in a pipe by Equation 8.4 is

$$E = \frac{Q^2}{2gA^2} + d$$

in which d is the depth of flow. This may be rewritten to express conditions for partially full pipe by writing

$$a = \frac{\text{actual flow area}}{\text{area of pipe}}$$

Introducing the discharge factor, which is equivalent to the Froude number,

$$q_c = \frac{Q}{D^{5/2}} \qquad (9.13)$$

and dividing both sides by the diameter D, the specific energy equation may be written in English units as

$$\frac{H_0}{D} = \frac{0.0252 q_c^2}{a^2} + \frac{d}{D}$$

which may be plotted in a dimensionless manner in Figure 9.25.

The most common materials used for culverts are concrete and corrugated steel. The roughness in both cases is usually assumed to be constant for any flow depth. Common values used in practice are $n = 0.012$ for concrete pipe and $n = 0.024$ for corrugated steel pipe. Although the selection of culvert materials is made in practice on a competitive economic basis, the fact that the roughness of corrugated steel pipes is twice that of concrete pipes suggests that their selection be based on hydraulic considerations also. On hilly terrains where the culvert slope is expected to be relatively steep and the flow through the culvert gains considerable energy, corrugated steel pipes offer energy dissipating advantages. On flat terrains energy loss through a culvert is undesirable; hence, concrete pipes are most suitable.

In Figure 9.23 the maximum discharge

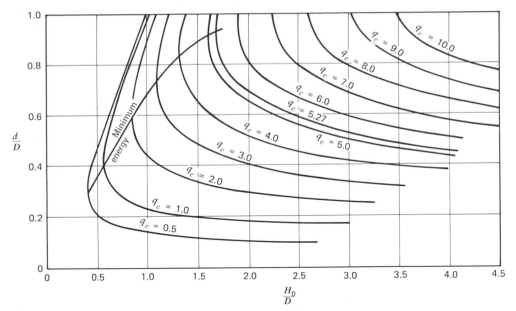

FIGURE 9.25 Specific energy graphs for circular pipes flowing partially full. $H_E = v^2/2g + d$ and $H_E/D = 0.0252(q_c^2/a^2) + d/D$, where $q_c = Q/D^{5/2}$; $a = A/A_{full}$.

in a pipe flowing under free surface conditions occurs at a depth that is 93% of the pipe diameter. For the given energy conditions the *optimum discharge* is found at the critical depth which is 93% of the diameter. On the vertical coordinate of Figure 9.24 this value gives $q_c = 5.27$ at the minimum energy line. Consequently by Equation 9.13 the optimum discharge in a culvert is

$$Q_{optimum} = 5.27D^{5/2} \qquad (9.14)$$

Substituting this value into Equation 8.22 the value of the *optimum slope* of a culvert may be determined as

$$S_{optimum} = 111\frac{n^2}{\sqrt[3]{D}} \qquad (9.15)$$

For the commonly used roughness values for concrete and corrugated steel pipes this equation may be plotted in the form of Figure 9.26. As long as the depth of the flow in the culvert is less than critical the structure will operate under inlet control. If the depth reaches its critical value, outlet control will take over. To aid the

computation of this problem the U.S. Bureau of Public Roads developed a series of graphs, one of which is reproduced in Figure 9.27. In this graph the dashed lines indicate the limiting outlet control conditions; the solid lines indicate the inlet control conditions for the culvert diameters indicated. For example the graph indicates that a 54 in. diameter corrugated metal pipe will always work under inlet control as long as its length L is equal or less than 50 $(100S_0)$ in feet. For longer lengths it will operate under outlet control, providing that the same culvert slope S_0 is retained.

When the tailwater depth exceeds the top of the exit pipe the culvert will be flowing full. Similarly, with efficient entrance geometry a culvert may flow under pressure, at least in the initial portion of the pipe. Flow conditions for culverts flowing full may be determined by the nomograph shown in Figure 9.28. It is a typical one of the several such nomographs made by the U.S. Bureau of Public Roads.

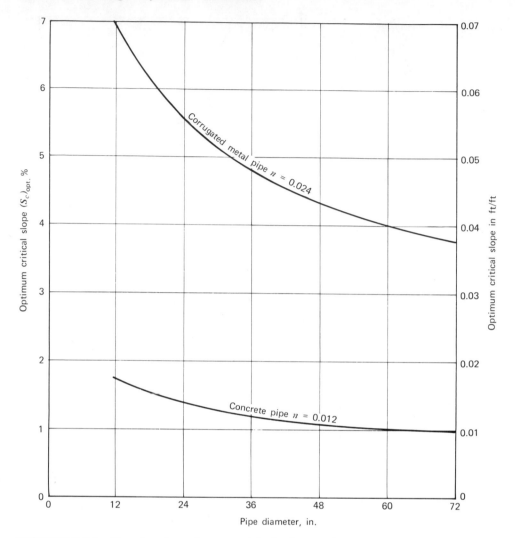

FIGURE 9.26   Graph of optimum critical slope of concrete and corrugated pipe culverts. Optimum conditions: $d_c/D = 0.93$; $Q/D^{5/2} = 5.27$. (From *Handbook of Concrete Culvert Pipe Hydraulics*, Portland Cement Association, 1964.)

At the exit point of a culvert the velocity of the water is considerable, because culverts are designed to operate near critical velocity. As a result it is often necessary to provide for energy dissipators at the outlet.

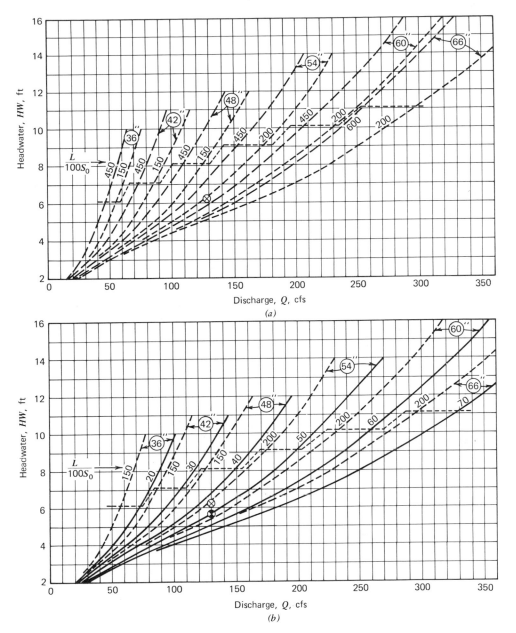

FIGURE 9.27 Typical culvert capacity chart. Standard circular corrugated metal pipe headwall entrance 36 to 66 in. in diameter. For 25% paved culverts, reduce length to 0.75 of actual L to compute $L/100S_0$. (*a*) For values greater than $L/100S_0$.

    Example: ⊗ Given: 130 cfs; $HW = 6.2$ ft; $L = 120$ ft.; $S_0 = 0.025$ in. ⊗ Select 54 in. unpaved $HW = 5.6$ ft.

    (From *Handbook of Concrete Culvert Pipe Hydraulics*, Portland Cement Association, 1964.)

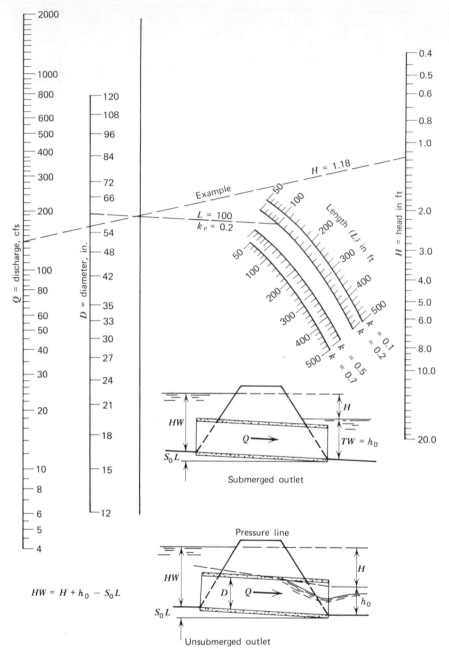

FIGURE 9.28 Typical nomograph for culverts under outlet control. (From *Handbook of Concrete Culvert Pipe Hydraulics*, Portland Cement Association, 1964.) Head for concrete pipe culverts flowing full; $n = 0.012$. Adapted from Bureau of Public Roads chart 1051.1.

Equation: $H = \left[\dfrac{2.5204(1 + k_e)}{D^4} + \dfrac{466.18 n^2 L}{D^{16/3}} \right] \left(\dfrac{Q}{10} \right)^2$

H = head in ft.
k_e = entrance loss coefficient
D = diameter of pipe in ft.
n = Kutter's roughness coefficient
L = length of culvert in ft.
Q = design discharge in cfs.

230

EXAMPLE 9.16

A 60 in. diameter concrete culvert is 100 ft long and its entrance loss coefficient is $k_e = 0.2$ (Figure E9.16). Determine the head loss in the culvert if the discharge is 140 cfs and the pipe is flowing such that both ends are submerged.

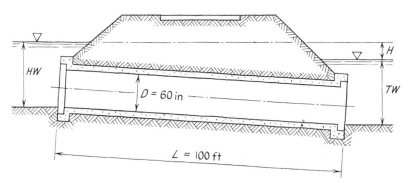

FIGURE E9.16

SOLUTION:

The culvert is flowing full. Assume the concrete roughness $n = 0.012$:

$$k_e = 0.2$$
$$L = 100 \text{ ft}$$
$$Q = 140 \text{ cfs}$$
$$D = \frac{60}{12} = 5 \text{ ft}$$

From the formula shown in Figure 9.28,

$$H = \left[\frac{2.5204(1 + k_e)}{D^4} + \frac{466.18n^2 L}{D^{16/3}} \right] (Q/10)^2$$

$$= \left[\frac{2.5204(1.2)}{5^4} + \frac{466.18(0.012)^2 \times 100}{5^{16/3}} \right] \left(\frac{140}{10} \right)^2$$

$$= [0.0048 + 0.0013]196$$

$$= 1.187 \text{ ft}$$

or from nomograph Figure 9.28, we find $H = 1.18 \text{ ft}$.

EXAMPLE 9.17

A corrugated metal culvert with a head wall entrance has a diameter of 48 in., and is 150 ft in length; it is laid on a slope of $S_0 = 0.02$ (Figure E9.17). For a head water elevation of $HW = 6$ ft, determine the discharge in the culvert and state whether the flow is under inlet or outlet control.

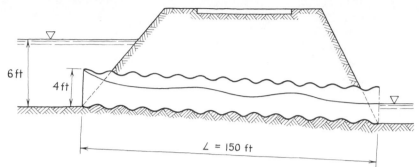

FIGURE E9.17

SOLUTION:

For corrugated steel culvert,

$$n = 0.024$$
$$S_0 = 0.02$$

Assume it is partially full, check by Equations 9.14 and 9.15

$$Q \text{ optimum} = 5.27 D^{5/2} = 5.27(4)^{5/2} = 168.64 \text{ cfs}$$

$$S \text{ optimum} = \frac{111 n^2}{D^{1/3}} = \frac{111(0.024)^2}{(4)^{1/3}} = 0.0402$$

Therefore, the culvert slope $S_0 < (S)_{\text{opt}}$

Check $50(100 S_0) = 50(100 \times 0.02) = 100 \text{ ft} < 150 \text{ ft}$. So the culvert is operated under outlet control. By employing Figure 9.27, $L S_0/100 = 150 \times 0.02/100 = 75$ feet and HW 6 ft, 48 in. diameter pipe we obtain the discharge $Q = 112 \text{ cfs}$.

COMPLEX STRUCTURES— MODEL ANALYSIS

Consider a circular, "morning glory" spillway built into a small reservoir and connecting to the downstream side of an earth dam through a simple sluiceway. The arrangement is shown in Figure 9.29.

Depending on the water level over the spillway crest and on the diameter of the spillway, the water may or may not enter into the vertical pipe under free surface conditions. Conversely, making reference to the previous discussion of culverts, the horizontal pipe may or may not operate under outlet control. To analyze a structure of this configuration no design formulas, neither graphs or nomographs may aid the designer. Yet the safety of the whole project may depend on the proper sizing of this common outlet structure. The only feasible way to obtain the necessary information on the operational characteristics of structures for which no data is available from a broad range of similar past experiences is by building a suitable model and testing it. Based on these test results, the model may then be adjusted by changing some of its parameters until the most desirable mode of operation is determined. For cost considerations models are rarely the size of the proposed structure. They are scaled down, retaining geometrical similarity, to a size that allows their cheap construction and their operation in a hydraulic labora-

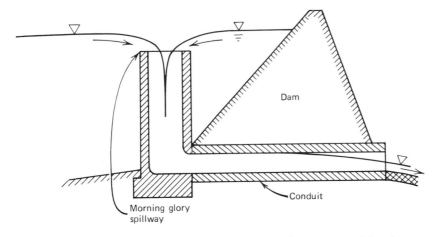

FIGURE 9.29 Morning glory spillway—sluiceway combination.

tory. The cost of such an experiment is minimal when compared to the cost of the structure, and infinitesimal when compared to the benefits of the results. Very often the savings, caused by a more rational design suggested on the basis of laboratory model studies, more than pay for the cost of these studies.

Hydraulic models of a full scale "prototype" are designed such that the motion and forces in them are similar to those in the prototype. To attain this the designer should be familiar with the concept and laws of *geometric*, *kinematic*, and *dynamic similarity*.

Geometric similarity means that all corresponding lengths in model and prototype are of the same ratio. Geometric scale of a model is selected so that it can be built and operated conveniently in a hydraulic laboratory and still big enough to measure velocities in it with sufficient accuracy. Model scales of spillways, conduits, and similar structures range between 1:15 and 1:50. Models of rivers and harbors range from 1:100 to as small as 1:2000. With such small scales water depths in the model would be so small that effects of surface tension and roughness could be overbearing. In such cases the vertical scale may differ from the horizontal scale and we speak of distorted models.

To satisfy the fundamental requirement that the model is to operate in a manner similar to the prototype one has to recognize that in each the flow follows Newton's law of motion, Equation 1.1. For the two systems

$$\left.\begin{array}{l} F_p = m_p \cdot a_p \\ F_m = m_m \cdot a_m \end{array}\right\} \qquad (9.16)$$

where subscripts p and m represent prototype and model. The acceleration is the time rate of velocity change

$$a = \frac{\Delta V}{\Delta T} \qquad (9.17)$$

and the mass is

$$m = \rho \cdot (L)^3 \qquad (9.18)$$

where L is the unit length, corresponding to either the prototype, L_p, or the model, L_m. To have geometric similarity between the two systems the ratio of lengths must be constant everywhere; hence

$$L_R = \frac{L_m}{L_p} \qquad (9.19)$$

This is called the model's scaling ratio.

Models corresponding to the prototype represented must operate in kinematic similarity. This means that the corresponding velocities in prototype and model must be constant throughout, that is,

$$V_R = \frac{V_m}{V_p} = \frac{L_m/T_m}{L_p/T_p} = \frac{L_R}{T_R} \qquad (9.20)$$

where T_R is the time ratio between times in the prototype T_p and times in the model, T_m. Equation 9.20 suggests that to have kinematic similarity

$$T_R = \frac{T_m}{T_p} \qquad (9.21)$$

must be maintained.

Returning our attention to Equations 9.16 and considering the succeeding formulas we may conclude that to have dynamic similarity between model and prototype, the ratio between corresponding forces should be

$$F_R = \frac{F_m}{F_p} = \frac{m_m}{m_p} \cdot \frac{a_m}{a_p}$$

$$= \frac{\rho_m}{\rho_p} \cdot \left(\frac{L_m}{L_p}\right)^3 \cdot \Delta\left[\frac{L_R}{T_R}\right] \Big/ \Delta T_R$$

$$= \rho_R (L_R)^3 \frac{L_R}{T_R^2}$$

that is,

$$F_R = \rho_R (L_R)^2 (V_R)^2 \qquad (9.22)$$

in which ρ_R is the ratio of fluid densities in the prototype and model, and

$$(L_R)^2 = A_R \qquad (9.23)$$

representing the ratio of the corresponding areas in model and prototype. The fundamental equation to assure dynamic similarity between prototype and model is, therefore,

$$F_R = \rho_R A_R V_R^2 \qquad (9.24)$$

In practical hydraulic structure design problems we are aware that the flow is influenced by a variety of forces. Inertia, gravity, viscous shear, capillarity, and elasticity all play a part. Usually the force of gravity or viscous shear forces are the most dominant. Because of the combined effect of the various forces acting in the prototype it is practically impossible to design a hydraulic model that would be a true representative of the prototype for all forces present. It is therefore necessary to judiciously neglect forces of lesser importance and concentrate one's efforts to satisfy modeling requirements for the most important forces. For only one dominant force, neglecting all other forces, one should be able to build a model at a selected scale and operate it in such a manner that the effect of nondominant forces are negligible. In this way model studies can give reasonably accurate information on the expected hydraulic behavior of the prototype. If there are only two types of forces considered, Equation 9.24 may be satisfied by selecting a model fluid different from water, which is the prototype fluid.

Most hydraulic models utilize water as a model fluid. Hence, the density ratio in Equation 9.24 is unity. In some cases other fluids may be used to obtain better dynamic similarity. Properties of common fluids that may be used in laboratory work are listed in Table 9-4. Once the density ratio is fixed Equation 9.24 will be a function of two independent variables only. These are L_R and T_R. One provides for geometric and the other for kinematic similarity.

Most hydraulic models either represent open channel flow, spillways, and weirs, for example, or represent pipe and closed conduit flow. The former are characterized by the work of gravity forces; the latter characterized by viscous forces. In Chapter 8 the Froude number was introduced (Equation 8.11) as a dimensionless parameter. It is the ratio of the inertia forces to the gravity forces. In Chapter 4 the Reynolds number was introduced (Equation 4.4) to represent the dimensionless ratio of the inertia forces to the viscous forces.

To assure that the Froude numbers in model and prototype are the same, one may write the ratio of gravity forces as

$$\frac{\gamma_m}{\gamma_p} \cdot \left(\frac{L_m}{L_p}\right)^3 = \gamma_R (L_R)^3 \qquad (9.25)$$

Since the Froude number is the ratio of gravity forces to inertia forces (and by

Table 9-4 Properties of Various Liquids Used in Hydraulic Modeling

	Temperature (degrees Fahrenheit)	Specific Weight		Density		Viscosity	
		lbs/ft^3	dynes/ cm^3	slugs/ ft^3	grams/ cm^3	lb-sec/ ft^2	dyne-sec/ cm^2
Ethyl alcohol	68	49.3	774	1.530	0.789	0.0000249	0.0119
Benzene	68	54.9	862	1.705	0.879	0.0000136	0.0065
Carbon desulphide	68	78.7	1237	2.446	1.261	0.0000077	0.0037
Carbon tetrachloride	68	99.5	1564	3.095	1.595	0.0000200	0.0096
Glycerin	68	78.7	1236	2.445	1.260	0.0313	14.99
Mercury	68	845.3	13,280	26.282	13.546	0.0000328	0.0157
Oil, castor	68	60.0	941	1.862	0.960	0.0206	9.86
Oil, linseed	60	58.3	916	1.812	0.934	0.000292	0.331
Oil, olive	59	57.3	900	1.781	0.918	0.0023	1.10
Oil, sperm	60	54.9	863	1.707	0.880	0.000877	0.42
Turpentine	61	54.5	856	1.694	0.873	0.0000334	0.016
Water, fresh	60	62.4	980	1.938	0.999	0.0000236	0.0113
Water, sea	60	64.0	1005	1.989	1.025	—	—

Equation 1.2) the dynamic similarity formula, Equation 9.24 should be equated with Equation 9.25 because their ratio must be unity. This results in

$$\gamma_R (L_R)^3 = \frac{\gamma_R}{g_R} (L_r)^3 \frac{L_R}{T_R^2}$$

from where it follows that

$$T_R = \sqrt{\frac{L_R}{g_R}}$$

if the Froude numbers in model and prototype are to be similar. Since the model is under the same gravitational acceleration as the prototype, g_R is always unity. Hence for models of open channels and free surface flow structures the time ratio will take the form

$$T_R = \sqrt{L_R} \qquad (9.26)$$

and the velocity ratio

$$V_R = \sqrt{L_R} \qquad (9.27)$$

Equations 9.26 and 9.27 represent the *Froude model law for time scale.*

When the flows are dominated by viscous shear, the model should operate such that the Reynolds numbers in the model should correspond to that in the prototype. Writing the ratio of viscous forces in model and prototype as

$$\mu_R \frac{(L_R)^2}{T_R} \qquad (9.28)$$

by Equation 1.3 we may equate Equation 9.28 with Equation 9.24 since their ratio must be unity, that is,

$$\rho_R \frac{(L_R)^4}{(T_R)^2} = \mu_R \frac{(L_R)^2}{(T_R)}$$

From this we may express T_R after simplification as

$$T_R = (L_R)^2 \frac{\rho_R}{\mu_R} \qquad (9.29)$$

which is the *Reynolds model law for time scale.*

From the two main model laws expressed by Equations 9.26 and 9.29 all other hydraulic parameters of dynamically similar models can be derived. For example, let us express the ratio of corresponding accelerations in model and prototype designed to satisfy the Reynolds model law.

$$a_R = \frac{a_{\text{model}}}{a_{\text{prototype}}} = \frac{L_R}{(T_R)^2}$$

By Equation 9.29 T_R is substituted, and

$$a_R = \frac{L_R}{\left[(L_R)^2 \dfrac{\rho_R}{\mu_R}\right]^2} = \frac{L_R \mu_R^{\,2}}{(L_R)^4 \rho_R^{\,2}}$$

Hence any acceleration measured in the prototype corresponds to model accelerations as

$$a_{\text{model}} = \left[\frac{\mu_R^{\,2}}{\rho_R^{\,2}(L_R)^3}\right] \cdot a_{\text{prototype}}$$

This formula may be referred to as the *transfer formula* between model and prototype. For the two main model laws these transfer formulas were determined for a number of hydraulic parameters that may be measured in prototype and model. The list of transfer formulas is shown in Table 9-5.

The roughness of a model built either under the Froude or the Reynolds model law presents another design problem. The scaling ratio of the n roughness coefficient between model and prototype is determined by

$$n_R = (g_R)^{-1/2}(L_R)^{1/6} \qquad (9.30)$$

For most models in which roughness may be a controlling factor the proper model roughness can be obtained by a trial and error procedure that should be carried out before the actual experimentation on the model begins. This involves observations of prototype behavior and adjustment of model roughness until the time ratio of the model and prototype correspond to its desired value. River and harbor models are the most sensitive to these effects; hence their geometric distortion is usually controlled by their relative roughness.

Table 9-5 Transfer Formulas for Froude and Reynolds Modeling Laws

Quantity	Modeling Ratios for:	
	Froude Law	Reynolds Law
Length	L	L
Area	L^2	L^2
Volume	L^3	L^3
Mass	$L^3 \gamma g^{-1}$	$L^3 \gamma g^{-1}$
Density	γg^{-1}	γg^{-1}
Time	$L^{0.5} g^{-0.5}$	$L^2 \nu^{-1}$
Velocity	$L^{0.5} g^{0.5}$	$L^{-1} \nu$
Acceleration	g	$L^{-3} \nu^2$
Angular velocity	$l^{-0.5} g^{0.5}$	$L^{-2} \nu$
Angular acceleration	$L^{-1} g$	$L^{-4} \nu^2$
Force and weight	$L^3 \gamma$	$\nu^2 \gamma g^{-1}$
Pressure	$L \gamma$	$L^{-2} \nu^2 \gamma g^{-1}$
Impulse and momentum	$L^{3.5} \gamma g^{-0.5}$	$L^2 \nu \gamma g^{-1}$
Discharge, volume/sec	$L^{2.5} g^{0.5}$	$L \nu$
Discharge, weight/sec	$L^{2.5} \gamma g^{0.5}$	$L^{-2} \nu^3 \gamma g^{-1}$
Energy and work	$L^4 \gamma$	$L \nu^2 \gamma g^{-1}$
Power	$L^{3.5} \gamma g^{0.5}$	$L^{-1} \nu^3 \gamma g^{-1}$
Torque	$L^4 \gamma$	$L \nu^2 \gamma g^{-1}$
Absolute viscosity	$L^{1.5} \gamma g^{-0.5}$	$\nu \gamma g^{-1}$
Kinematic viscosity	$L^{1.5} g^{0.5}$	ν
Surface tension	$L^2 \gamma$	$L^{-1} \nu^2 \gamma g^{-1}$
Modulus of elasticity	$L \gamma$	$L^{-2} \nu^2 w \gamma g^{-1}$

EXAMPLE 9.18

For the structure described in Example 9.17 design a hydraulic model for which the geometric ratio is 1/10. Develop transfer formulas for discharge and roughness such that the model will truly represent the prototype in question.

SOLUTION:

Since the geometric ratio is $L_m/L_p = 1/10$, $L_R = 0.10$. It is clear that the gravity force is dominant force, so the model is to satisfy Froude's model law. Since from Problem 9.17,

$$L_p = 150 \text{ ft}$$
$$n = 0.024$$
$$s_0 = 0.02$$
$$H_p = 6 \text{ ft}$$
$$Q_p = 112 \text{ cfs}$$

we have $L_m = L_p \cdot L_R = 150 \times 0.1 = 15$ ft (Equation 9.18) for the length of our model. The diameter of the model culvert pipe is similarly computed as

$$L_m = 4 \text{ ft} (L_R) = 4(0.1) = 0.4 = 4.8 \text{ in.}$$

The operating upstream head 6 ft is 0.6 ft or 7.2 in.

To satisfy the dynamic similarity conditions we note that since $g_R = 1$, then from Equation 9.26 $T_r = \sqrt{L_R}$. Thus the discharge transfer formula is

$$Q_R = A_R V_R = (L_R)^2 \frac{(L_R)}{T_R} = \frac{L_R^3}{L_R^{1/2}} = L_R^{2.5} = 0.1^{2.5} = 0.0032$$

Since
$$Q_R = \frac{Q_{\text{model}}}{Q_{\text{prototype}}}$$

the model discharge will be

$$Q_{\text{model}} = 0.0032 Q_{\text{prototype}}$$

This means, for example, that if the optimum discharge in the prototype is $Q_p = 168.64$ cfs, in the model similar conditions will be attained at 0.533 cfs, assuming that all other modeling characteristics are satisfied. The most important of these is the proper modeling of the pipe roughness. From $n = 1.49 R^{2/3} S^{1/2}/V$ where S is dimensionless, then

$$n_R = \frac{n_m}{n_p} = \frac{[R^{2/3} S^{1/2}]_m}{[R^{2/3} S^{1/2}]_p} \cdot \frac{v_p}{v_m} = \frac{L_R^{2/3}}{V_R} = \frac{L_R^{2/3}}{L_R/T_r}$$

and by introducing the time scale of Froude's law,

$$n_R = \frac{\sqrt{L_R} \, L_R^{2/3}}{\sqrt{g_R} \, L_R} = \frac{L_R^{1/6}}{g_R^{1/2}}$$

which is Equation 9.30. Since $g_R = 1$ and $L_R = 0.1$ then $n_R = (0.1)^{1/6} = 0.68$.

Since the roughness of the prototype was stated in Example 9.17 to be

$$n_p = 0.024$$

then the roughness in the model should be

$$n_{\text{model}} = 0.68(n_{\text{prototype}}) = 0.68(0.024)$$
$$= 0.0163$$

Checking in Table 8-2 showing roughness coefficients, we note that the most commonly used model pipe material, lucite, has a roughness coefficient of 0.009, which is almost half the roughness factor required for the model. Therefore, the pipe in the model should be artificially roughened to attain the required conditions.

EXAMPLE 9.19

For the broad crested weir flow problem described in Example 9.6 design a hydraulic model of a scale $L_R = 1/5$ and write the transfer formulas for all major variables. The purpose of the model study is to determine proper discharge coefficients.

SOLUTION:

From Example 9.6

$$Q_p = 476 \, \text{cfs}$$
$$b_p = 30 \, \text{ft}$$
$$h_p = 2 \, \text{ft}$$

Hence

$$h_m = \frac{2}{5} = 0.4 \, \text{ft} = 4.8 \, \text{in.}$$

Since the flow is essentially two dimensional, the width of the spillway may be reduced at will as long as side friction in the model is negligible. Usually the width of the available laboratory flume will establish the value of b. Assuming that the flume width available is 3 ft, the discharge is to be adjusted accordingly. It is advisable, to avoid confusion, to use the discharge per unit width, q.

$$q = \frac{Q}{b} = \frac{476}{30} = 15.86 \text{ for the prototype}$$

Since the problem is controlled by gravity flow, Froude's model law will apply with $L_R = 0.2$.

From Table 9-5 the transfer formulas are

$$\text{Discharge} = L_R^{2.5} \cdot g_R^{0.5} = 0.0179$$
$$\text{Energy} = L_R^4 \cdot \gamma_R = 0.0016\gamma_R$$
$$\text{Viscosity} = L_R^{1.5} \cdot \gamma_R \cdot g_R^{-0.5} = 0.089\gamma_R$$

Since $g_R = 1$, its functions will not enter the problem. Also, since both model and prototype will use the same fluid, water, γ_R will be unity.

To get the transfer equations we have

$$Q_{model} = 0.0179(Q_{prototype}) = 0.0179(15.86)$$
$$= 0.284 \text{ cfs/ft of prototype}$$

or

$$Q_{model} = (0.284)0.2 \times 3 = 0.17 \text{ cfs}$$

for the 3 ft wide modeling flume.

We now recall that the head h_1 over the weir was introduced as a measure of the driving energy; hence, it is to be modeled as such. As long as the model fluid is water the modeling ratio for the head is as given above for energy, that is,

$$h_{model} = 0.0016 h_{prototype}$$
$$= 0.0032 \text{ ft}$$
$$= 0.038 \text{ in.}$$

which is an extremely small quantity, subject to forces due to surface tension and other factors.

Disregarding the energy consideration, by simply using the geometric scaling ratio, we would get

$$h_{model} = 0.2(2) = 0.4 \text{ ft} = 4.8 \text{ in.}$$

Otherwise, we may consider using another fluid, in which case since

$$\gamma_{prototype} = 62.4 \text{ lb/ft}^3$$

and for the energy transfer formula

$$E_R = \frac{h_{model}}{h_{prototype}} \text{ to be equal } L_R = 0.2$$

$$E_R = 0.2 = 0.0016 \frac{\gamma_{model}}{\gamma_{prototype}} = 0.0016 \frac{\gamma_{model}}{62.4}$$

$$\gamma_{model} = \frac{0.2(62.4)}{0.0016} = 7800 \frac{\text{lb}}{\text{ft}^3}$$

an improbably heavy model fluid to find.

To further complicate the problem, we may consider the discharge coefficient to be some way dependent on the viscosity. In that case the ratio of the viscosities of the fluids in the model and prototype would be defined by

$$\mu_R = \frac{\mu_{model}}{\mu_{prototype}} = 0.089 \gamma_R$$

And since the unit weight and viscosity of the prototype water is known, we would get

$$\mu_{model} = 0.089 \frac{\gamma_{model}}{62.4} (\mu_{prototype})$$

and if $\mu_{prototype}$ equals one centipoise, then

$$\mu_{model} = 0.0014 \gamma_{model}$$

the γ_{model} previously was determined as 7800; hence,

$$\mu_{model} = 11.125 \text{ centipoises}$$

which is an exceedingly high value for model fluids.

Accordingly, in order to perform this simple model experiment in an absolutely correct manner, one would either have to develop a yet unknown and improbable model fluid, or move to another planet where the gravitational acceleration is markedly different. The other alternative is to relax our stringent requirements for modeling and accept a degree of uncertainty about the behavior of the model. This shortcoming may be overcome by full scale proof testing and comparing the two results.

GENERAL DESIGN CONSIDERATIONS

In the foregoing discussions we reviewed the basic methods to determine the fundamental hydraulic dimensions of structures. From these data the designer proceeds to design the size of the hydraulic structure itself. In many cases the hydraulic designer is aided by experts in structural engineering and soil mechanics. The basic design data, however, is supplied by the hydraulic designer. These, in addition to hydraulic dimensions, include hydrostatic pressures from upstream as well as downstream seepage considerations and the associated uplift forces due to groundwater, erosion, and many other considerations. The designer should keep in mind considerations related to the possibility of dewatering the structure in case of maintenance or repair, possible catastrophic occurrences damaging the structure, the required corrective actions, and the ways that they may be carried out.

The most common errors made by inexperienced designers of hydraulic structures are as follows:

Underestimating the flows expected during the life of the structure. To avoid this, care should be taken in the proper hydrologic design.

Undersizing the structural parts of the structure resulting in erosion, underwashing, and the subsequent gradual collapse of critical structural portions. Sizes of headwalls, spillway aprons, wingwalls, and other components subject to scour, earth pressure, seepage, erosion, and other effects should be in many instances two or three times larger than suggested by the designer inexperienced in hydraulic structure design.

Failing to consider floating debris that may be carried by the water during critical floods is another common error that should not be blamed on legal "acts of God."

In certain geographic areas the effect of ice should be considered. This necessitates the understanding of the thermodynamic fundamentals of water introduced in Chapter 1.

ICE FORMATION AND CONTROL

Ice forms when the water is supercooled even if only by a few fractions of a degree below the freezing point. To change from liquid to solid phase the water must expel its latent heat of crystallization to its surroundings. In still water the recipient of this heat is the atmosphere. Hence in still water freezing starts at the surface. If the water is in motion turbulent mixing allows a transfer of heat from the lower layers. Since the specific heat of the solid bank and bottom material is less than that of the water, they may cool faster. Water next to these materials on the bottom and at the edges of the surface where the water velocity is small will freeze if supercooled. Steel is very conducive for underwater ice formation, concrete is less so, while little ice may form on wood under water. Microscopic sediment particles act as "seed" for freezing. By picking up the latent heat in supercooled water they become centers, nuclei, of frozen minute droplets. Following this, neighboring particles of water freeze also and, like a cloud, collect in a spongelike mass. At low velocities the lower density of ice causes this "slush ice" to rise to the bottom of the surface ice layer increasing its thickness. In faster flows the small ice particles are carried in suspension by the random turbulent velocities.

Depending on the mechanics of its formation we speak of surface ice and subsurface ice. The latter may be divided into bottom or anchor ice, and frazil or slush ice. On some rivers the mass of subsurface ice may be two to three times that of surface ice. The combined effect of surface ice and subsurface ice may reduce the hydraulic radius of the flow channel (Equation 8.17) by as much as 80%. During sustained periods of regional freezing the remaining flow section may be enough to carry the flow. In case of a sudden warming the frozen channel will prove to be insufficient to carry the flood. The rising waters will cause the ice cover to break up. The fractured surface ice will be carried downstream. The floating ice may collect into ice jams on the outward side of stream bends or where the channel is constricted. Bridge piers, gates, and spillways are susceptible to such blockage. These occurrences have been known to cause major disasters. The horizontal pressure of the piled up ice layers is so large that they may damage the structure or prevent the raising of vertical gates. In the mid-fifties gates of a dam in Czechoslovakia had to be blasted away within a year of their completion because they became jammed under ice. To enable gates to be raised they are often installed 20 to 30 degrees off vertical so that when they are raised they pull away from the ice upstream. Floating booms or permanent skimmer walls are often built over the intake canals of hydraulic power stations to prevent surface ice from entering the turbines.

Blasting was found to be the most effective means of removing major ice jams from open streams. To prevent the ice loosened by blasting from freezing together again, blasting of an ice jam should always proceed from the downstream side of the block toward the upstream in V-shaped strips of 100 ft or so, with the point upward and over the deepest point of the channel as shown in Figure 9.30. For each strip blasting starts by first opening up a series of holes in the ice sheet along the

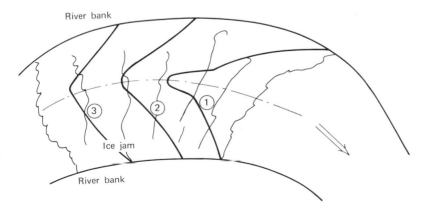

FIGURE 9.30   Firing line pattern for ice jam removal.

planned firing pattern. The main charges will then be lowered under the ice through these holes. The holes may be made by explosive charges placed on the surface of the ice. To blast a hole of the required size, either conventional or special shaped explosive may be used. A 0.5 kg charge of dynamite placed on the surface will blast a hole in a 0.5 m thick layer of ice. An M3 shaped military charge from an average 42 in. standoff creates a 12 ft deep hole with a diameter of 6 in. These holes, if placed about 5 m from each other, will be sufficient to crack the ice mass if sufficiently charged and fired simultaneously. One of the most common mistakes is to underload the charges. To break up the ice the required amount of medium strength dynamite, per hole, is computed by

$$E = 0.6t^3 \qquad (9.31)$$

where E is the explosive charge in kilograms and t is the thickness of the ice in meters. This amount, if placed about 2 m below the ice surface, will crack up a hole of about 3 to 5 m in diameter. Military experience shows that an 80 lb charge of M3 demolition blocks under $4\frac{1}{2}$ feet of ice creates a crater 40 ft in diameter. The charge in each hole is lowered on a rope and allowed to float under the ice a short distance, as shown in Figure 9.31. The rope is supported by a wooden bar holding it across the hole. It is advisable to pre-

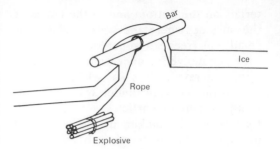

FIGURE 9.31 Arrangement of explosive charge in preblasted hole in ice.

pare all charges before the blasting operation commences, since the weakened ice jam may be too dangerous to work on later. For most effective jam removal charges in each row of holes must be exploded simultaneously using electric fuses.

Subsurface ice presents various operating difficulties for hydraulic structures. The steel frame and guide structure of gates, valves of sluiceways, and intake structures may allow the formation of "anchor ice," hindering the operation of these structures. For larger structures this may be prevented by electric heaters built into the structure. Screens and racks are also very susceptible for clogging by slush ice. In many locations complete removal of racks and screens is necessary before the cold season begins.

PROBLEMS

9.1 An 8 ft long 18 in. diameter rough steel drain pipe with a sharp entrance is opened to allow flow under a concrete dam under a 10 ft depth of water. How much will be the discharge through the pipe?

9.2 A 4 m wide rectangular sluice gate is opened to a height of 1 m. The upstream water level is 4 m, at the downstream it is 2 m. Calculate the discharge under the gate.

9.3 What will be discharged under the gate described in Problem 9.2 if the downstream water level is 3.5 m?

9.4 A 2 miles long flood wall with a 2 ft crown width is overtopped by a flood. The depth of the water at the crown is 0.5 ft. Estimate the amount of flow waters entering the protected side during a period of 6 hours.

9.5   A rectangular shaped 2 ft thick broad crested weir is 6 ft long. The height of the water upstream from the weir is 1 ft above the crest, at the downstream it is 6 in. above the crest. Determine the type of flow across the weir and the discharge flowing if the crest height is 4 ft.

9.6   For the situation described in Problem 9.5 compute the shape of the overfalling stream of water if the downstream water level is 4 ft below the crest.

9.7   A 48 in. diameter culvert is flowing half full. The discharge is 200 cfs. Determine the total energy in the flow.

9.8   What is the optimum discharge in a 6 ft diameter culvert?

9.9   Compute the optimum slope of an 8 ft diameter corrugated metal culvert.

9.10   Determine the time scale for a model built to represent the prototype described in Problem 9.1. Assume that water will be used in the model. How long will be the required time to drain the water from a model reservoir built at a geometric scale of 1/10 if the drainage time of the prototype is 24 hours?

10

FLOW MEASUREMENTS

Whether in the field or in the laboratory, discharge and other measurements provide the fundamental data on which hydraulic analysis and design is based. This chapter contains a broad survey of the most important hydraulic measuring devices that are currently available for the hydraulician. The reduction and presentation of hydraulic data concludes the chapter.

Measuring discharges and other flow parameters are an essential part of the analysis and operational control of all hydraulic systems. The rate of use of water by municipal customers, industrial processes, and agricultural users is usually metered. The capacity of watercourses and hydraulic structures must be determined by direct measurements. Hydrologic study of creeks and rivers is based on the statistical analysis of a long sequence of data obtained by repeated or continuous measurements. For these various applications a multitude of methods and devices were developed over the years. All of these are based on the fundamental physical laws of fluid mechanics discussed in Chapter 2.

CLASSIFICATION OF FLOW MEASUREMENTS

Basically flow determinations may be made directly and indirectly. *Direct* deter-mination of the discharge could be made by volumetric or weight measurements. The simplest way of measuring a discharge is to allow the flow into a container of known volume and measuring the time required to fill it. In laboratories containers placed on scales allow the weighing of the water entering the container in a measured time interval. The measured weight, divided by the unit weight of water and by the measured time interval, gives the flow rate in volume per unit time. Municipal water meters measure the volume of water used over a period by such direct measurements. These are called positive displacement meters as they measure the volume of water that fills a given container repeatedly. Usually the number of fillings is counted by a register or counter and the total quantity is obtained as the product of the volume of a chamber and the number of fillings. The registers are geared such that the counters are showing the total volume in gallons, cubic feet, or liters. Hence the displacement meters

244

may be looked upon as fluid motors operating at a high volumetric efficiency under a very light load. The types of these water meters include the oscillating or rotary piston meters, sliding or rotating vane meters, nutating disc meters, gear or lobed impeller meters, and other devices named according to their mechanical design. In addition to their importance in the customer's water use determination, which is of little interest to the hydraulic designer, these meters that register positive displacement were introduced recently in a broad range of hydraulic discharge measurements. By replacing head loss measuring and registering devices in orifice or flume type hydraulic measuring structures, they serve as proportional secondary circuit meters. This type of application will be discussed later as appropriate.

Indirect determination of the flow involves defining or establishing known flow conditions and measuring one or more parameters, such as pressure or its variation, kinetic energy, and water surface elevations. The measured parameters along with the hydraulic formulas applicable define the flow rate. According to the appropriate laws of fluid mechanics the various techniques and devices may be classified by their practical applications. By this approach one may group the flow measuring methods as they apply to:

Pipe networks
Rivers
Creeks and other small watercourses
Hydraulic structures
Hydraulic laboratories

In the following the various flow measuring techniques will be discussed in the sequence listed above.

PIPE FLOW MEASUREMENTS

Indirect flow measurements in pipelines are made, in practice, by a great variety of commercially available measuring devices. All of them use the principle of continuity and one other equation which defines the motion. The latter most commonly is the energy equation (Equation 2.9). Orifices, measuring nozzles and venturi tubes, Pitot tubes, Prandtl tubes, and Annubars® operate on this principle. Other devices use the momentum equation (Equation 2.5). These include the propeller meters and elbow meters. The viscous drag force along with the impulse force are measured by the Ramapo® Mark V device. Other physical principles used in flow measurement include Faraday's law of electromagnetic induction, used in magnetic flow meters, and Ohm's law, utilized in hot wire anemometers measuring the rate of cooling by the flowing fluid of an electrically heated resistance wire. Several other new measuring concepts are under development.

The simplest method of measuring discharge in a pipe is offered by a *circular orifice*. Its arrangement is shown in Figure 10.1. By virtue of the equation of continuity one may observe that

$$Q = (V \cdot A)_{\text{pipe}} = (v \cdot a)_{\text{orifice}} \quad (10.1)$$

If we introduce the energy equation with the assumption that the pipe is horizontal, Equation 2.9 takes this form:

$$\frac{V_p^2}{2g} + \frac{p_p}{\gamma} = \frac{v_0^2}{2g} + \frac{p_0}{\gamma} + h_2 \quad (10.2)$$

where h_2 is an energy loss occurring at the orifice edges and in the turbulent exit zone behind the orifice plate. For orifices and similar flow meters the energy loss is expressed as a coefficient c, which is experimentally recorded at the time of calibration. Combining Equations 10.1 and 10.2 and rearranging the terms, one may write the ideal average velocity through an orifice as

$$v = \sqrt{\frac{(2g/\gamma)(p_p - p_0)}{1 - [c(a/A)]^2}} \quad (10.3)$$

and the discharge Q as

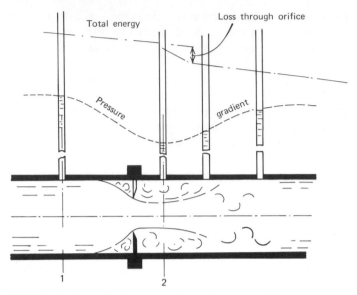

FIGURE 10.1 Flow through a circular orifice.

$$Q = \frac{ca}{\sqrt{1 - c^2(d/D)^4}} \sqrt{\frac{2g}{\gamma} \Delta p} \quad (10.4)$$

where d and D are the diameters of the orifice and the pipe, respectively, and

$$\Delta p = p_p - p_0 \quad (10.5)$$

Experimental studies indicated that the actual value of the c coefficient depends on the location where p_0 is measured. In addition it also depends on the configuration of the edge of the orifice, on the ratio d/D, and on the Reynolds number in the pipe. For Reynolds numbers above 25,000 the value of the c coefficient becomes a constant for all diameter ratios below 0.5. Since these are representative of practical cases, a good approximate formula for orifices in practical applications may be written in the simple form of

$$Q = ca\sqrt{\frac{2g}{\gamma} \Delta p} \quad (10.6)$$

The value of c is to be determined by calibration. Commercially available orifice meters are supplied with a calibration chart. The discharge coefficient is usually within the range of 0.6 to 0.7. For water at constant temperature and in fully turbul-ent flow, the discharge equation of a given orifice can be represented by the simple formula of

$$Q = K \Delta p^{1/2} \quad (10.7)$$

where K is a parameter lumping all variables that are fixed for the orifice in question.

To reduce the losses occurring in the constricted flow and to extend the range of the constant values of the coefficient, orifices are sometimes substituted by nozzles or venturi tubes as shown in Figure 10.2.

The derivation of the discharge formula for *flow nozzles* or *venturi tubes* is identical to that for the orifice meter. Discharge coefficients of these devices range from 0.94 to 0.98.

In metering any fluid in a pipeline, it is most important that the flow approach the orifice, flow nozzle, venturi, or any other metering element in a normally turbulent state. The flow must not be influenced by swirls, cross-current, eddies, or other disturbances that create helical paths of flow. Disturbances after the flow measuring element may introduce errors into the static pressure measurements at that point. To

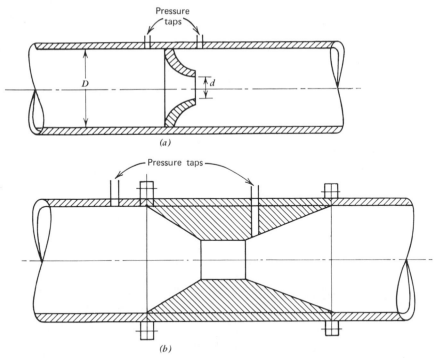

FIGURE 10.2   (*a*) Flow nozzle and (*b*) venturi tube.

produce uniform conditions it is recommended that at least a length of six to ten
times the pipe diameter be of straight pipe
in front of measuring devices. If this is not
possible in some installations, then guide
vanes are to be installed in the pipe to
straighten the flow. After the metering
device a straight length of three to five
times the pipe diameter is desirable.

EXAMPLE 10.1

A 12 in. diameter pipe is equipped with a 4 in. diameter orifice
plate (Figure E10.1). By calibration the discharge coefficient was
determined to be 0.63. What is the discharge in the pipe if the
differential pressure registered during the flow equals 12 psi?

SOLUTION:

From Equation 10.6

$$Q = ca\sqrt{\frac{2g}{\gamma}(\Delta p)}$$

where

$$c = 0.63$$

$$a = \frac{\pi}{4}\left[\frac{4}{12}\right]^2 = 0.0872 \text{ ft}^2$$

$$\gamma = 62.4 \frac{\text{lb}}{\text{ft}^3}$$

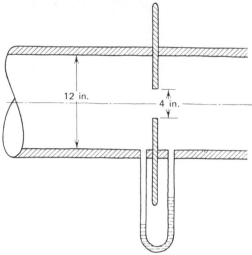

FIGURE E10.1

$$\Delta p = 12 \times 144 = 1728 \frac{\text{lb}}{\text{ft}^2}$$

Then

$$Q = 0.63 \times 0.0872 \sqrt{\frac{2(32.2)(1728)}{62.4}}$$

$$= 2.32 \text{ cfs}$$

EXAMPLE 10.2

A flow nozzle of 6 in. diameter is placed in an 18 in. diameter pipe. During calibration the pressure differential was measured to be 10 psi for a flow of 7.2 cfs and 17 psi for a flow of 9.3 cfs. Determine the discharge coefficient of the measuring nozzle (Figure E10.2).

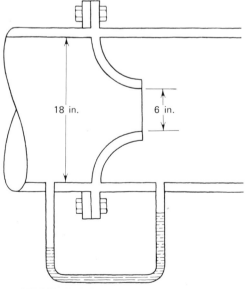

FIGURE E10.2

SOLUTION:

From Equation 10.6,

$$Q = ca\sqrt{\frac{2g}{\gamma}(\Delta p)}$$

$$a = \frac{\pi}{4}\left(\frac{1}{2}\right)^2 = \frac{\pi}{16}\,\text{ft}^2$$

$$c_1 = \frac{7.2}{\frac{\pi}{16}\sqrt{\frac{2\times 32.2}{62.4}\times 10\times 144}}$$

$$= 0.95$$

$$c_2 = \frac{9.3}{\frac{\pi}{16}\sqrt{\frac{2\times 32.2}{62.4}\times 17.0\times 144}}$$

$$= 0.942$$

$$c = \tfrac{1}{2}(c_1 + c_2) = \tfrac{1}{2}(0.95 + 0.942) \simeq 0.946$$

EXAMPLE 10.3

An orifice meter has a 2 in. orifice plate and is placed in an 8 in. diameter pipe. The meter is to be placed into a straight pipe of 8 ft length. Where should the meter be located?

SOLUTION:

The most efficient spot to place the meter is where the flow is uniform and undisturbed. To locate it properly at least 6 to 10 times of the pipe diameter from the inlet flow is recommended. As we have 8 ft of pipe, 10×8 in. $= 80$ in. $\simeq 6$ ft from the inlet point. In this case 2 ft remains between the meter and the pipe, which equals 3 diameters. This is still sufficient to eliminate downstream influences.

The *measurement of pressures* and pressure differences is performed by manometers, differential manometers, or Bourdon gages. The simplest method to measure pressure in a closed pipe is by the use of a transparent vertical or slanted standpipe, called *piezometer tube*, connected to a tap on the pipe in which the pressure is to be measured. By Equation 3.1 the pressure energy in the water will force the water level up to a height y, at which the static pressure in the piezometric tube equals that in the pipe itself, as shown in Figure 10.3. Piezometric tubes are used for some laboratory work, but for practical applications they are not feasible. Since the height required is proportional to the unit weight of the fluid in the piezometer, large pressures could be measured more conveniently by using mercury that is 13.5 times heavier than water. To prevent the mercury from flowing out into the pipe the piezometer is bent into a U-shaped traplike loop. The resulting device is called a *manometer* (Figure 10.4).

The simple manometer shown in Figure 10.4a allows the determination of the pressure in the pipe at point m by writing

$$p_m = \gamma_{Hg}z - \gamma h \qquad (10.8)$$

in which γ_{Hg} is the unit weight of mercury

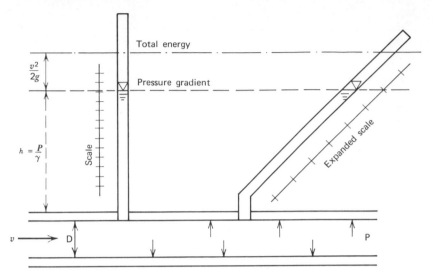

FIGURE 10.3   Piezometric tubes for pressure measurement.

and γ is that of water. For a mercury manometer, therefore, Equation 10.8 takes the form of

$$p_m = \gamma(13.5z - h) \qquad (10.9)$$

In this case we assumed that the pressure at point b is atmospheric; hence, p_m is the pressure above atmospheric pressure. If the other side of the manometer is connected to another pressure tap, as shown in Figure 10.4b we speak of a differential manometer. A *differential manometer* is used to measure differences in pressure. Because of differences in pressure at the points m and n, there exists a difference in level, z, between the two mercury columns in the U tube. By starting at m where the pressure is p_m, and noting the changes in pressure as we pass to points c, d, and n, the following expression for p_n is obtained:

$$p_m - \gamma x - \gamma_{Hg}z + \gamma y = p_n$$

or

$$p_m - p_n = \Delta p = \gamma(x - y) + \gamma_{Hg}z$$

Referring to Figure 10.4b, one may see that

$$x - y = -h - z$$

and by substitution

$$\Delta p = \gamma_{Hg}z - \gamma z - \gamma h$$
$$= \gamma[z(13.5 - 1) - h] \qquad (10.10)$$

In case the two pressure taps are at the same level, then $h = 0$ and

$$\Delta p = 12.5\gamma z \qquad (10.11)$$

in case mercury is used as manometer fluid. For other manometer indicating fluids, Equation 10.10 must be rewritten accordingly.

When the pressure difference between two measured points is very small, the elevation difference in adjacent piezometer pipes is difficult to determine. It is convenient in such cases to magnify the difference by using a manometer fluid that is lighter than water. Various manometer oils are available commercially having unit weights about 80% that of water. Differential manometers designed to work with these lighter than water manometer fluids are of the inverted kind, as shown in Figure 10.4c. There is a variety of manometer fluids used within the range of mercury and light oils. Usually these are color-coded such that their specific gravity may be ascertained beyond doubt. One manufacturer's manometer fluids are listed in Table 10-1. In the use of manome-

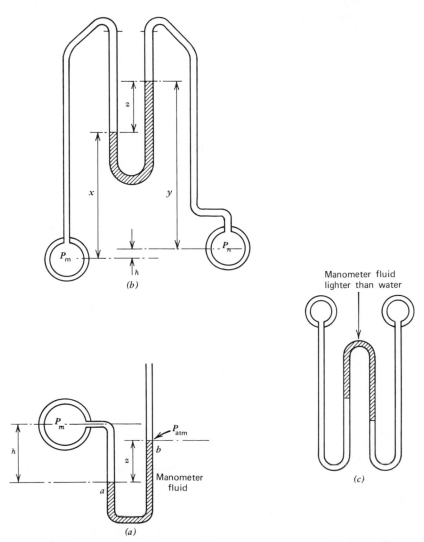

FIGURE 10.4   Pressure determination by manometers. (*a*) Simple manometer, (*b*) differential manometer, and (*c*) inverted manometer.

ters care must be taken to expell all air bubbles from all connecting tubes, otherwise the readings obtained will have little value. For correct measurements the pressure taps in the pipes must be free of burrs and the entrance holes should be on a surface parallel to the velocity in the pipe.

Table 10-1 Manometer Fluid Properties

Trade Name	Color	Unit Weight g/cm³	Description
Meriam Red Oil	Red	0.827 at 60°F	A highly refined mineral seal oil. Non-corrosive. For use where a light fluid is required and extremes are not encountered. Useful temperature range is 40 to 120°F.
Meriam Indicating Fluid Concentrate	Green	1.000	Specially prepared concentrate for high precision work. Noncorrosive to brass, or stainless steel. Low surface tension. Hysteresis practically zero. Keeps indicating tube clean. Not for use under freezing conditions.
Meriam Unity Oil	Red	1.00 at 73°F	A mineral seal oil mixture prepared for general use from 30 to 100°F, where a fluid near the specific gravity of water is desired and water is unsuitable.
Meriam Indicating Fluid	Red	1.04 to 83°F	Specially suited to high vacuum applications. Teflon gasketing recommended. Noncorrosive. Insoluble in water. Useful temperature range is 20 to 150°F.
Meriam Indicating Fluid	Light straw	1.20 at 77°F	A specially prepared oil particularly suited for general outdoor work. At elevated temperatures, corrosive to brass and aluminum. Useful range is 10 to 150°F.
Meriam Indicating Fluid	Blue	1.75 at 55°F	Extremely stable. For use in manometers and flowmeters. Low viscosity, nonfreezing, nonflash. Insoluble and clear interface in water. Useful range is 70 to 150°F.
Meriam No. 3 Fluid	Red	2.95 plus or minus 0.002 at 78°F	A heavy bromide. Nonflashing. All parts in contact with this fluid must be brass, glass, or stainless steel. Corrosive to steel. Useful temperature range is 40 to 100°F.
Meriam Hi-Purity Mercury	Silver	13.54 at 73°F	Specially treated for highest purity. Used in all applications where greatest density is required for maximum range and where chemical reaction is not encountered. Cannot be used in aluminum or brass instruments.

EXAMPLE 10.4

The maximum working length (scale length) of a manometer is 16 in. Calculate the maximum pressure differential that may be measured if the indicating fluid is Meriam Blue.

SOLUTION:

From Table 10-1 we find that the unit weight of Meriam Blue indicating fluid is 1.75 g/cm³.

Assuming that the pressure taps are at identical level, h in Equation 10.10 is zero. Rewriting Equation 10.10 for $\gamma_{indicator} = 1.75$, we have

$$\Delta p = \gamma z (1.75 - 1) = \gamma z (0.75)$$

Observing z on Figure 10.4, we note that

$$z_{maximum} = l = 16 \text{ in.}$$

if l is the scale length of the manometer. Therefore, if the pressure is expressed in the form of a height of a water column h, then

$$h_{max} = \left(\frac{\Delta p}{\gamma}\right)_{max} = 0.75(16) = 12 \text{ in.}$$

EXAMPLE 10.5

A manometer pressure differential in two adjacent $(h = 0)$ pipe lines is 15 psi. Select a manometer fluid for a differential manometer such that the length of the manometer does not exceed 8 in.

SOLUTION:

From Equation 10.10

$$\Delta P = \gamma_m z - \gamma z$$
$$= \gamma z (m - 1)$$
$$15 \times 144 = 62.4 \times 8(m - 1)$$
$$m - 1 = 4.33$$
$$m = 5.33$$

where m = specific gravity of manometer fluid so the fluid to be used has to have a unit weight larger than

$$\gamma = 5.33 \frac{g}{cm^3}$$

Measurement of high pressure may also be made by ordinary steam gages invented by Bourdon. *Bourdon gages* consist of a coiled tube having its inner end closed and connected by a simple rack and pinion to a hand that is free to rotate over a graduated dial. The pressure in the coiled tube tends to uncoil the tube, which results in a movement of the hand. The dial is calibrated by applying a known pressure to the tube and marking the position of the hand on the dial as shown in Figure 10.5.

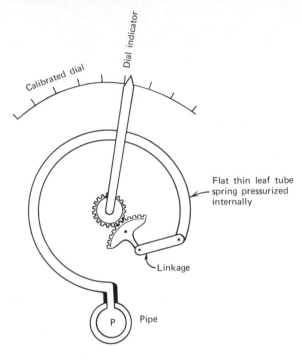

FIGURE 10.5   Bourdon gage.

Modern technology developed a variety of pressure-sensing devices to be used in fluids. The most common types measure the deflection of a circular elastic membrane, the diaphragm, which is under pressure. The deflection is transmitted to a gage or meter either mechanically, magnetically, or electrically.

Magnetic linkage between diaphragm and dial pointer in *Dwyer Magnehelic®* gages is provided by a cantilever leaf spring that is located behind the dia-

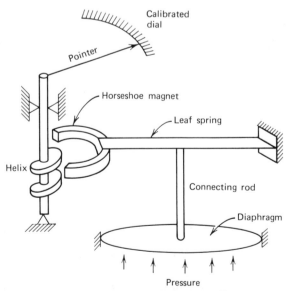

FIGURE 10.6   Dwyer Magnehelic® gage.

phragm and mechanically connected to it. Deflection of the diaphragm by fluid pressure behind it bends the leaf spring. At the end of the leaf spring a horseshoe magnet is affixed. Between the ends of this magnet there is a circular staircase-shaped helix made of a metal of high magnetic susceptibility. Since the helix is pivoted at its end, it rotates freely moving a pointer at one of its ends. As the horseshoe magnet moves, the helix tends to maintain a minimum gap between the ends of the magnet. Hence the change of pressure on the diaphragm causes the pointer to move, indicating the pressure, as shown in Figure 10.6.

Electrical linkage between the deflecting elastic membrane and a calibrated indicating meter or a recorder is provided by *pressure transducers*. Pressure transducers are usually built such that a variable resistance electric strain gage is bonded to the membrane. The strain measured by the strain gage results from differential fluid pressure on the membrane. Hence the change in measured voltage flowing through the gage is proportional to the pressure of the fluid on the diaphragm. Changing of pressure range to be measured can be done by simply changing the diaphragm. Another type of electric pressure transducer involves the principle of capacitance. A thin stretched stainless steel diaphragm welded to its case forms a variable capacitance with an insulated electrode located very close to the diaphragm. The pressure change induced variation in capacitance is

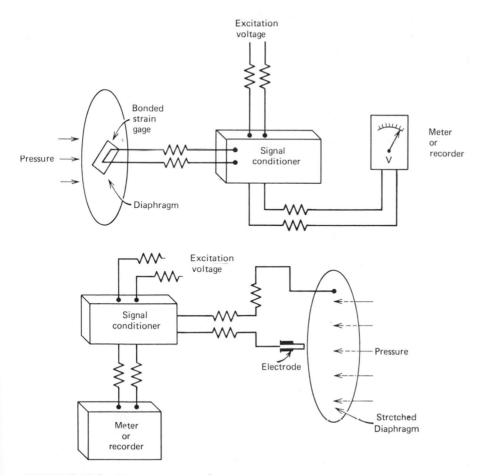

FIGURE 10.7 Pressure transducers.

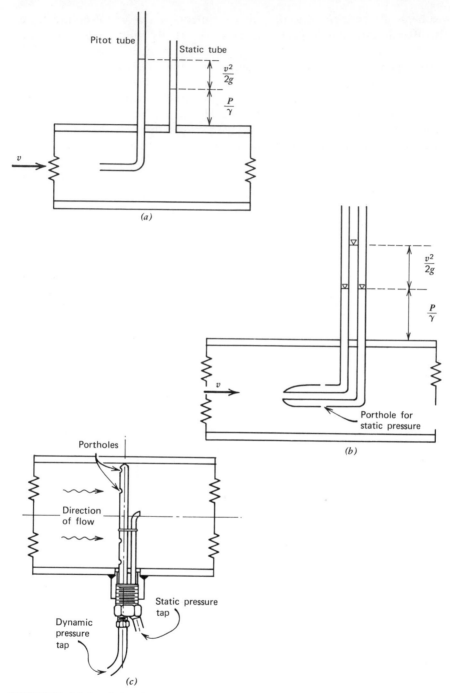

FIGURE 10.8 (a) Pitot tube, (b) Prandtl tube, and (c) Annubar®.

translated by electronic means into high level linear direct current signal. Gages made by Setra Systems, Inc., operate on this principle. Schematic drawings of these devices are shown in Figure 10.7.

The advantages of electric pressure transducers in practice are based on the facts that such signals can be amplified to attain greater precision, that they can be continuously recorded by magnetic or strip chart recorders, and that electric signals can be transmitted for remote control applications.

Velocity measurements in pipes, as well as open channels, may be made by measuring the pressure corresponding to the kinetic energy of the flow. The simplest method is to measure the pressure in an open-ended tube bent in such a way that the end is aligned opposite to the velocity vector measured. Such tubes are called Pitot tubes. The kinetic energy at the center of a Pitot tube converts into pressure energy as the flow of the fluid is halted at that point, as shown in Figure 10.8a. In pipes the pressure differential h between a Pitot tube and a nearby static tube provides the means to compute the velocity at the point measured, such that

$$v = \sqrt{2g\,\Delta h}$$

in which $\Delta h = \Delta p/\gamma$. Hence

$$v = \sqrt{\frac{2g}{\gamma}\,\Delta p} \qquad (10.12)$$

By moving the Pitot tube across the diameter of the pipe the velocity distribution may be obtained from which the discharge can be calculated. Fixed Pitot tubes may be calibrated for a range of discharge to allow the discharge determination directly by multiplying the measured pressure differential by a constant.

The combination of static and Pitot tubes resulted in the Prandtl tube shown in Figure 10.8b. Prandtl tubes are available commercially in various sizes. One, made by the United Sensor and Control Corporation is as small as 1/16 in. in diameter.

EXAMPLE 10.6

Pressure differential measured on a Prandtl tube inserted into a flowing stream was 6.5 in. of water (Figure E10.6). What is the velocity of the flow?

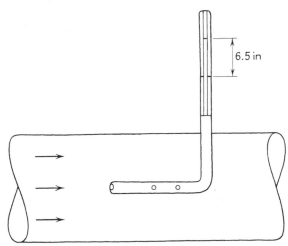

FIGURE E10.6

SOLUTION:

By Equation 10.12

$$v = \sqrt{2g\,\Delta h}$$

where
$$\Delta h = 6.5 \text{ in.} = 0.54 \text{ ft}$$
Therefore
$$v = \sqrt{32.2 \times 2 \times 0.54}$$
$$= 5.9 \text{ fps}$$

Further improvement of this type of measurement is represented by the *An-nubar*® shown in Figure 10.8*c*. Kinetic energy measured in the Annubar is a composite of the velocities in several annuli of the pipe cross section. These measuring devices are calibrated in the factory and they give the discharge of the flow directly, by the use of a calibration curve.

Utilizing the impulse-momentum principle, velocity measurements in pipes may also be made by *propeller meters*, although most of these meters are used in rivers and open channels. They are used for velocity measurements in hydraulic power stations, in which case several propeller meters are attached to a cross bar located in the conduit leading to the turbine. For very small discharges (0.15 gallons per hour to 6 gallons per minute) a notable new propeller meter for pipe flow applications was developed by the Bearingless Flowmeter Company. In this, as its name implies, a small rotor spins freely without any contacts with the meter housing. The bearingless flow meter, shown in Figure 10.9, is equipped with an optical readout. The rotor's light-reflective marks are sensed by a photo-detector through fiber optics. The photo-detector produces an electrical impulse for each rotor mark passing a window and registers it by a counter.

Another flow measuring device utilizing the momentum-impulse principle is the *elbow meter*. In a pipe elbow the direction of the velocity changes and, by virtue of Equation 2.5, an impulse force is generated acting along the outer curvature of the bend. By measuring the difference in pressure between the inner and outer curvature of the pipe bend, as shown in Figure 10.10, the effect of the impulse force is registered. By proper calibration the discharge in a pipe elbow of a given

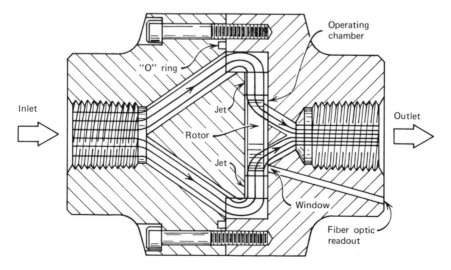

FIGURE 10.9　Bearingless flow meter. (Reproduced with permission from Bearingless Flow Meter Co., Boston, Mass.)

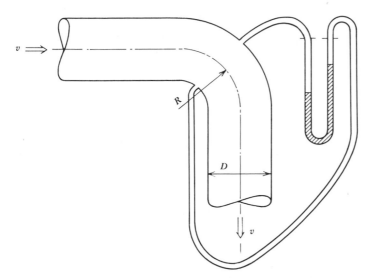

FIGURE 10.10   Pressure distribution in pipe elbows.

diameter may be determined. An approximate discharge formula for 90° pipe elbows for which the radius of curvature of the pipe R is larger than twice the pipe diameter may be expressed as

$$Q = \frac{\pi D^2}{4} \sqrt{\frac{R}{2D}} \sqrt{2g} \sqrt{\frac{\Delta p}{\gamma}}$$

$$= 0.785 \frac{D^{1.5} R^{0.5}}{\rho^{0.5}} \sqrt{\Delta p} \qquad (10.13)$$

where Δp represents the pressure difference between the inside and outside pressure taps located along the 45° diagonal of the elbow. For continuous discharge measurements pipe elbows may be equipped by a standard positive displacement water meter known from municipal metering practice. In such arrangements the difference in pressure between the taps will force some water to flow across the meter. This discharge, registered by the meter, will be proportional to the pressure difference, which, in turn, varies directly with the discharge in the pipe. The system, once calibrated, is a useful method to determine the output of pumping installations.

Another class of instruments measuring discharge in piping systems is the *variable area meter*. These include the tapered tube and float meter, the multiple-holed cylinder and piston, and the slotted cylinder and piston types.

In all variable area meters the energy causing the flow through the meter is substantially constant for all discharges and the discharge is directly proportional to the metering area. In the *tapered tube and float meter*, shown in Figure 10.11, the area variation caused by the upward flowing fluid produces a rise or fall of the float in the tapered tube. The float is designed so that the impulse force and viscous drag force raising it, less the force of buoyancy, are constant for all positions of the float in the tapered tube. These instruments are referred to as rotameters. In the Fisher-Porter laboratory rotameter the tapered tube is made of glass to make the position of the float visible. The discharge versus float elevation is determined by calibration. Since the viscous drag force lifting the float is dependent on the viscosity of the fluid, which in turn is dependent on the temperature, for different fluids or calibration fluid at different temperature the actual readings of the meter must be reduced to calibration data. The actual discharge q_a of a fluid with actual unit weight corresponds to calibration discharge q_c as

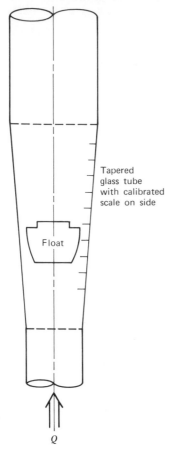

FIGURE 10.11   Rotameter.

$$q_a = q_c \sqrt{\frac{\gamma_c}{\gamma_a}} \qquad (10.14)$$

where γ_c is the unit weight of the fluid for which the meter was calibrated. The *slotted cylinder type variable area meter* utilizes a cylinder with a slot on its side through which the flow exits the meter. The varying rates of flow will cause a piston in the cylinder to rise or fall. The position of the piston is seen through a sight tube in front of a calibrated scale. An example of the slotted cylinder type meter, shown in Figure 10.12, is the MEM (Meter Equipment Manufacturing, Inc.) Series A flow meter.

With direct visual display the variable area flow meters give an instantaneous discharge value. The *Ramapo® Mark X flow meter* overcomes this shortcoming

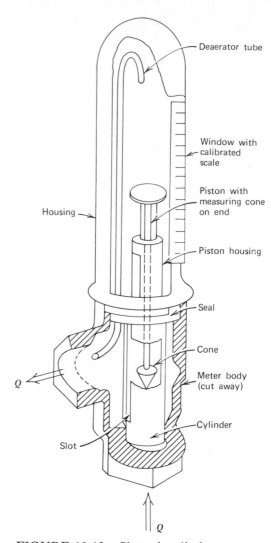

FIGURE 10.12   Slotted cylinder meter.

by transforming the float position in its tapered tube into an electric signal. This is accomplished by a linear variable differential transformer (LVDT) connected to the float by a vertical stem. The principle of an LVDT is shown in Figure 10.13. It consists of an electromagnet excited by a primary AC current. The position of the core element is dependent on the mechanical input—in the case of the flow meter it is connected to the float. The relative location of the core is sensed by two secondary coils by the induced currents generated in them according to Faraday's

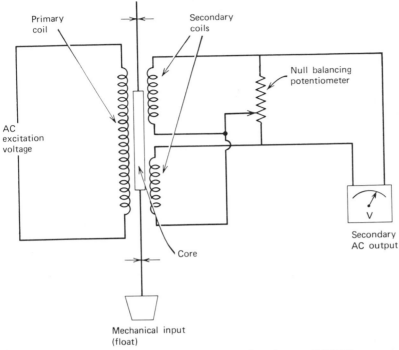

FIGURE 10.13 The principle of operation for an LVDT.

principle of electromagnetic induction. When the core is centrally located the electromagnetically induced currents in the two secondary coils are equal. If the core is offset, one secondary coil will have more induced current than the other one. By electric instrumentation the position of the core, hence, the connected measuring float, can be determined and recorded. The output voltage is calibrated to the discharge of the flow meter. Faraday's principle of magnetic induction is used directly for discharge measurement in *magnetic flow meters*. The principle, shown in Figure 10.14, involves a magnetic field generated by an electromagnet placed perpendicular to the pipe in which the flow takes place. As the fluid flows through the magnetic field an electromotive force is induced in it. The strength of this electromotive force is proportional to the velocity of the fluid and may be measured by a galvanometer connected to two electrodes attached to the pipe. Such instruments are manufactured by the Foxboro Company and others. The Series 600

Voltmeter, made by Cushing Engineering, Inc., is built with four electrodes and is hence capable of measuring two perpendicular velocity components simultaneously. This instrument is usable in open channel flows as well, in a velocity range of 0.3 to 30 ft/sec.

Other new pipe flow measuring devices include the *electrostatic flow meter*, based on two dissimilar metal pipe sections connected by a nonconducting pipe section. The voltage change between the two metals relates to the flow rate of the fluid. Also under development is the *thermal wave flow meter*, which correlates the time of flow of a periodic thermal wave between an electric heater element and a sensor.

RIVER FLOW MEASUREMENTS

Direct determination of river discharge is usually done by the concurrent measurements of depth and velocity distribu-

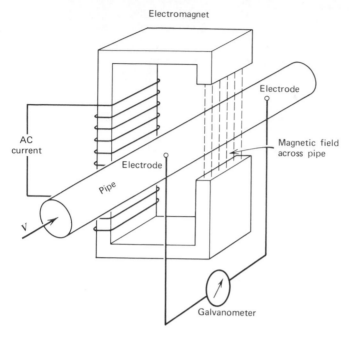

FIGURE 10.14   Magnetic flow meter.

tions at a suitable cross section of the stream. The best location to perform such measurements along a stream is where the direction of the velocity is substantially perpendicular to the section, cross currents are minimal, and the cross section is reasonably uniform and free from vegetable growth. Bridges are ideal locations for such measurements; otherwise the use of guy-cables and boats are required. Figure 10.15 shows the usual arrangement of such measurements.

In shallow rivers where the depth does not exceed three to four meters and the velocity is less than 0.5 m/s, graduated *sounding or wading rods* are used for measuring the depth. Deeper channels are measured by *cables with weights* at their end. The size of these weights depends on the maximum velocity expected in the stream. For velocities ranging up to 0.5 m/s a 5 kg weight is sufficient; for 2 m/s a 50 kilogram weight is necessary. For high velocities up to 4 m/s, 100 kg weights are recommended. To handle

these large weights the measurement is performed by utilizing motorized winches.

To account for the deflection of the cable due to the velocity of the water, the approximate formula

$$h = l \cos \alpha \qquad (10.15)$$

may be used in which the true depth is h, the length of the submerged cable is l and α is the initial angle of the cable with the vertical. This equation is reasonably valid as long as the initial angle is less than 30 degrees.

More precise measurements of depth may be performed by *ultrasonic devices*. These measure the time required for a sound pulse emitted by the instrument to bounce back from the bottom to a receiver in the manner sonar operates. *Echographs* plot the results of continuous measurements across the stream.

Measuring depth of water by *electric* means is also possible by the Model 100

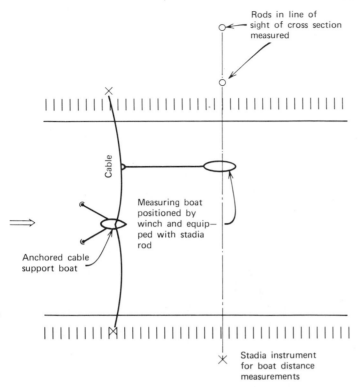

Rods in line of
sight of cross section
measured

Cable

Measuring boat
positioned by
winch and equip—
ped with stadia
rod

Anchored cable
support boat

Stadia instrument
for boat distance
measurements

FIGURE 10.15   River flow measurement from boats.

water level gage made by Marsh-McBirney, Inc. This device uses two stainless steel wires mounted parallel. One of these wires carries an electric current. The second wire senses the average voltage placed in the water by the first wire. The measured voltage is linearly proportional to the depth of immersion.

For a known channel cross section the velocity is measured along a number of verticals. For medium size rivers at least 10 to 15 verticals are measured and distributed evenly across the channel. At each vertical velocities are measured at several depths. For shallow depths one point measurement at 0.6y is sufficient. For depths up to 0.6 meter, two points measured at 0.2y and 0.8y are generally sufficient. In these cases the average velocity is determined by Equation 8.3. In three point measurements, at depths of 0.15y, 0.5y, and 0.85y, the average velocity is still the arithmetic mean of the three measured values. For uneven velocity distribution along a deep vertical the measurements are taken at five points. In addition to the measured velocity values at near the surface, v_s, and at near bottom, v_b, measurements are taken at 0.2y, 0.6y, and 0.8y. The average velocity in this case is computed by

$$v_{\text{ave}} = \frac{v_s + 3v_{0.2} + 2v_{0.6} + 3v_{0.8} + v_b}{10} \quad (10.16)$$

Results of discharge measurements in a river can be graphically represented in a plot similar to the one in Figure 10.16. The discharge Q may be determined from the data by computing the discharge q in each strip between two adjacent verticals. In this case the discharge of a strip between two verticals is

$$q = \left(\frac{v_1 + v_2}{2}\right)\left(\frac{h_1 + h_2}{2}\right) \cdot b \quad (10.17)$$

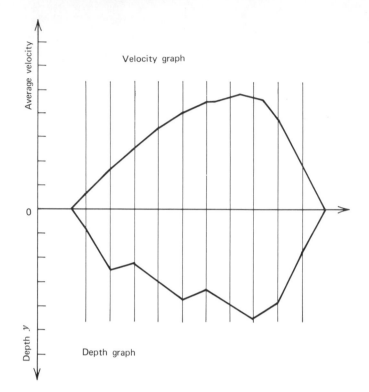

The table shown in the figure:

y	v_{ave}	Q
y_1	v_1	q_1
y_2	v_2	q_2
y_3	v_3	q_3
y_4	v_4	q_4

$$Q = \Sigma q$$

FIGURE 10.16   Representation of data from river flow measurement.

where b is the width between the adjacent verticals and subscripts 1 and 2 refer to the measurements taken in the verticals. The total discharge is then the sum of that at all strips.

A more precise result may be obtained by graphical solution. In this, the velocities are plotted on the cross section where they were measured and points of equal velocity are connected by contour lines. The areas of known velocities may be measured by a planimeter. The sum of all areas multiplied by their respective velocities give the total discharge of the stream.

Performing the discharge measurement of a river by the direct method described above is a time-consuming process. As the flow in a river is rarely constant but either rises or falls, the measured velocities must be adjusted to an average value, valid for the duration of the measurement. This adjustment must be carried out if the average depth of the stream changes by more than 1% during the measuring time. To establish the change of stage it is recommended that the water level be marked by a stake at the shore before and after the measurement, and that the change of average water level during this time be recorded.

The commonly used velocity measuring device in rivers is the *propeller meter* shown in Figure 10.17a. There are various propeller meters available commercially. A typical propeller meter consists of a propeller runner, a revolution counter, a shaft, and a tail piece. The propeller is usually seven to twelve centimeters in diameter. The pitch of the propeller is dependent on the average velocity for which the meter is designed. Normal propellers are designed to operate within the velocity range of 0.03 to 10 m/s. Care

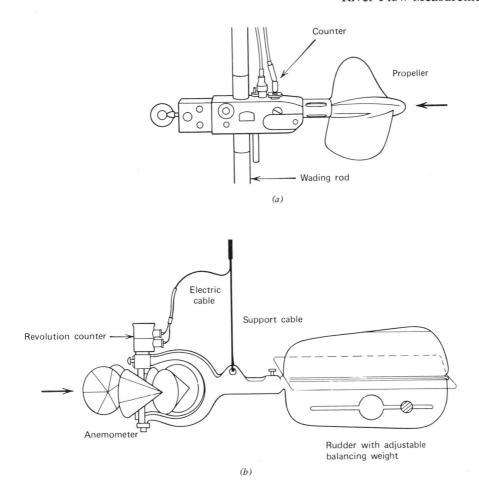

FIGURE 10.17   Current meters. (*a*) Propeller meter and (*b*) Price current meter.

is taken to develop propellers such that they react to velocity components in their axial direction only. This enables the operator to measure the true velocity passing through the river cross section perpendicularly, when the meter is held by a rigid rod. The counter gives an electric signal audible through an ear phone after each or after every 5, 10, or 20 revolutions. The tail piece acts as a rudder, holding the meter in the direction of the flow, and is used only if the current is known to be perpendicular to the cross section measured.

Of the various propeller meters the best known are the German Ott (Kempten), French Neypric, and the Soviet Zhestovsky. Their main difference is in their propeller design. Notable new development in this field is the Ott propeller meter in which the propeller turns a small electric generator and the velocity measured is proportional to its electric output. Another significant new device is the Hungarian VITUKI's* direction-indicating propeller meter, which is held through vertically and horizontally placed potentiometers, in a gyroscopic suspen-

* Research Institute for Water Resources, Budapest, Hungary.

sion system. By the specially designed propeller and rudder the meter takes the position of the maximum velocity. This orientation is indicated electrically by the aid of the potentiometers. Experience gained with this meter indicates that the directional error with some self-aligning velocity meters could be as much as 20%.

The *Price current meter* used in the American practice is manufactured by Teledyne-Gurley Company and, in contrast to propeller meters, uses a cup-type anemometer rotating around a vertical shaft as shown in Figure 10.17*b*. It is

manufactured in various sizes for different applications and may be used either on a wading rod or on a cable with a weight attached. The direct reading current meter includes a six pole permanent magnet and a reed switch sending electric signals to a direct readout unit on the surface, which shows the velocity of the water on a dial indicator.

All current meters, whether of propeller or anemometer configuration, operate on the momentum-impulse theory. To use them one must have a calibration table. Calibration of these instruments is done in

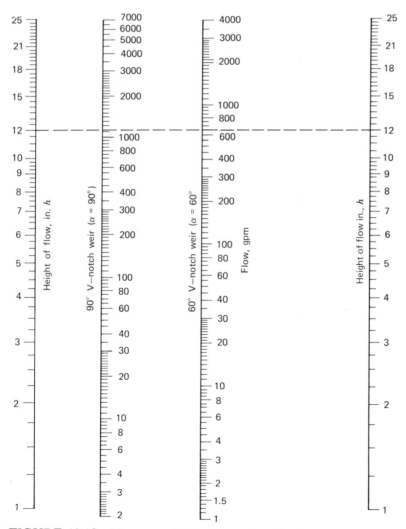

FIGURE 10.18   Nomograph for V-notch weir discharge. (Courtesy of Public Works Magazine.)

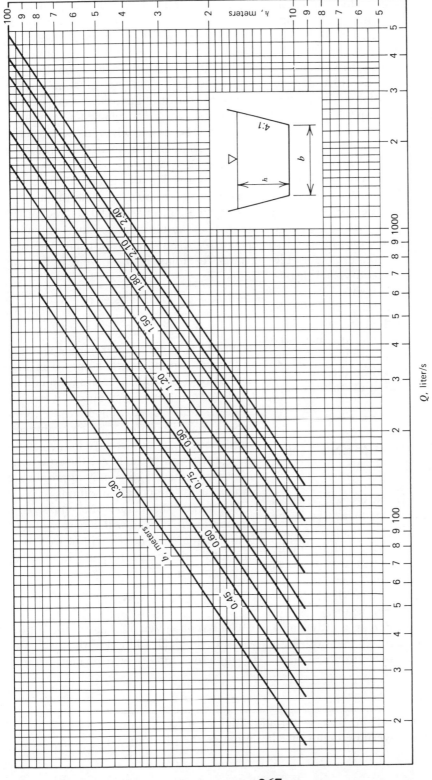

FIGURE 10.19 Discharge nomograph for Cipoletti weirs. (Courtesy of Public Works Magazine.)

267

towing tanks. The instruments are attached to a carriage that is pulled over the towing tank at various velocities during which the rates of revolution of the meters are recorded. Under careful maintenance and proper operation the attainable precision with propeller-type current meters ranges between five and 10%, depending on the rate of pulsation in the current. The larger the pulsation in the stream the greater the possible error.

Magnetic flow meters, mentioned earlier were also developed for measuring velocities in rivers, lakes, and estuaries, as well as sewers and flumes. The electromagnetic current meters made by Marsh-McBirney, Inc., measure two velocity components within the range of zero to 10 ft/sec. They feature portable-rechargeable batteries and visual display of both velocity components.

Average stream velocity can also be determined by the difference in the times of arrival of *ultrasonic pulses* measured by two sensors equidistantly located upstream and downstream on the shore opposite from the location of an ultrasonic sending device.

FLOW MEASUREMENT IN SMALL CREEKS

Shallow rocky rivers, creeks, or small open channels cannot be measured in the same manner as rivers because of their small size. In the absence of hydraulic structures that may be utilized as points of measurement either portable weirs or flumes are utilized or the discharge is determined by chemical means.

Portable weirs or flumes are available or may be constructed at the site. With proper calibration such structures may give a reliable information on the flow rate, but their shortcomings are significant. If the soil in which they are placed in the channel is pervious, the prevention of seepage around them is almost impossible. The requirement of free overfall makes it necessary to dam up the upstream water, resulting in considerable pressure on the upstream side of the structure, requiring significant structural supports. In cases when these portable weirs are applicable, their discharges are determined by the weir discharge formula, Equation 9.9. The discharge coefficient is generally determined by calibration.

For small discharges 90 degree or 60 degree V-notch weirs, the so-called *Thomson weirs* are often used. The discharge formulas for these weirs in English units are

$$Q_{90°} = 2.284h^{5/2}$$
$$Q_{60°} = 1.312h^{5/2} \qquad (10.18)$$

where Q is in gallons per minute and h is in inches. The nomograph shown in Figure 10.18 is useful in solving Equations (10.18). The above equations are valid for smooth, knife-edged weir plates. Experience shows that the V-notch weirs are very sensitive to any change in the roughness of the weir plate; hence the equations can be assumed to give only approximate values under practical conditions.

For larger discharges the trapezoidal *Cipoletti weir* is used with a base width of b and steep side slopes with 4 to 1 rise. Cipoletti's discharge formula in English units (gpm and in.) is

$$Q = 3.367bh^{3/2} \qquad (10.19)$$

The graph given in Figure 10.19 provides a graphical solution of the discharge formula for Cipoletti weirs in metric units.

EXAMPLE 10.7

A Cipoletti weir of 12 in. base width and sides of 4 to 1 slope carries water at an overflow depth of 8 in. What is the discharge measured?

SOLUTION:

From Equation 10.19,

$$Q = 3.367bh^{3/2}$$

where

$$b = 1 \text{ ft}$$

$$h = \frac{8}{12} \text{ ft}$$

Then

$$Q = 3.367 \times 1.0 \times \left(\frac{8}{12}\right)^{3/2}$$

$$= 4.124 \text{ cfs}$$

Chemical discharge measurements are most valuable for small streams where a large part of the flow passes through the rocks and gravel of the stream bottom. The common principle of the chemical discharge measuring techniques is based on the continuity equation. It involves the introduction of a known amount of tracer chemical into the water, which, after traveling in the stream over a certain length, will completely diffuse into the water. The amount in which the tracer chemical is found in the water after complete mixing is proportional to the discharge of the stream. Hence, by determining the degree of saturation of tracer in the water, its discharge may be computed. For complete mixing the recommended stream length, L, in metric units, is

$$L = 0.13C \frac{(0.7C - 6)}{g} \cdot \frac{B^2}{H}$$

(10.20)

where C is Chézy's coefficient for the stream given by Equation 8.19; B is the surface width and H is the depth of the stream.

The degree of saturation of the chemical can be determined by sampling and by proper testing procedures. Depending on the chemical used, these procedures may involve chemical, colorimetric, flame-photometric, radioactive, or electric conductivity measurements. Because of the relative simplicity of electric conductivity

measurement in the field, only the latter method will be discussed here. Since other methods may require the collection of considerable number of samples in the field, the analysis of these samples must be performed under laboratory conditions.

There are two ways to introduce tracer chemicals into the stream. One is the continuous introduction of a constant tracer discharge Q_t of a known degree of saturation C_t (in milligrams of tracer material per liter of water). The discharge of the stream Q then is determined by measuring the degree of saturation C of the tracer in the stream, and we write the equation of proportionality such that

$$Q = \frac{C_t}{C} Q_t$$

(10.21)

Another way of measuring the discharge is by a tracer wave. In this method a given amount of tracer material is mixed in a container of water, which then is dumped into the creek. At the point of sampling the degree of saturation is measured continuously or at frequent intervals (15 or 30 seconds) until the degree of saturation of the passing wave of tracer material becomes insignificant. The "wave" of the passing tracer at the sampling point may be plotted as shown in Figure 10.20 and the area under the time versus tracer saturation is determined. The area A under the curve, in the dimen-

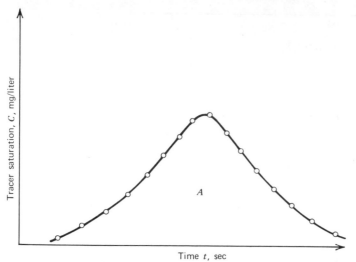

FIGURE 10.20 Discharge determination by tracer wave.

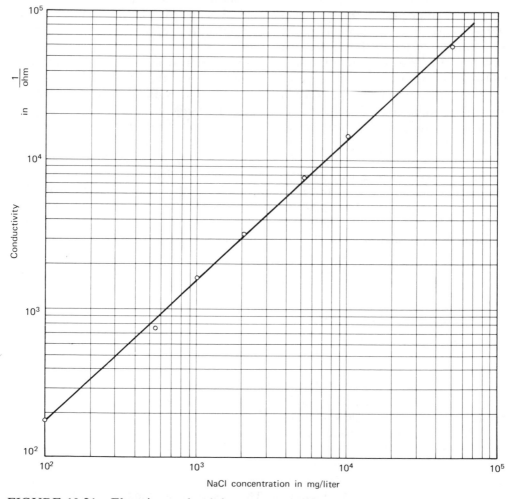

FIGURE 10.21 Electric conductivity versus salt content in water.

sions of second · mg/l, defines the discharge Q (liter/s) of the creek as

$$Q = \frac{E}{A} = \left[\frac{mg}{s \cdot mg/liter} \right] = \left[\frac{liter}{s} \right] \quad (10.22)$$

where E is the tracer input in milligrams. Of all materials used as tracers by far the most convenient for practical use is sodium chloride, the ordinary salt. This allows the use of commercially available portable *electric conductivity meters*. One of these, Model 33 made by Yellow Springs Instrument Company is particularly suited for tracer discharge measurement. It also measures temperature and dissolved oxygen in the water. Its advantage is that it does not require a continuous zero-balancing during measurements.

Salt's maximum degree of saturation is 240 g/liter; hence a 5 gallon container of water may dissolve a 10 lb bag of salt. The electric conductivity of salt solutions shown in Figure 10.21 allows for conversion from the readings on the conductivity meter into degrees of saturation. While salt is often found in nature this base amount can be eliminated from the procedure by starting the recording before the dumping of the tracer and ending the recording when the readings return to near the original base conductivity. This way the area under the tracer wave will exclude the base readings.

EXAMPLE 10.8

A salt wave measurement in a small creek was performed by dumping brine containing 9.8 lb of salt (4.45×10^5 mg NaCl) into a stream. At 200 ft downstream from the dumping point, a portable conductivity bridge read at each 30 sec gave the results as shown below with the corresponding salt contents shown in the third column as determined by Figure 9.21.

Time (sec)	$\frac{1}{ohm}(\times 10^2)$	mg/liter NaCl
30	5.1	3400
60	5.0	3333
90	5.0	3333
120	5.0	3333
150	5.0	3333
180	7.5	5000
210	10.0	6667
240	8.5	5660
270	7.5	5000
300	7.0	4660
360	6.5	4330
390	6.1	4070
420	6.0	4000
450	6.0	4000
480	5.8	3860
510	5.5	3660
540	5.6	3730
570	5.4	3600
600	5.4	3600

Cont'd

Time (sec)	$\dfrac{1}{ohm}(\times 10^2)$	mg/liter NaCl
630	5.4	3600
660	5.4	3600
690	5.3	3530
720	5.2	3460
750	5.1	3400
780	5.0	3333
810	5.2	3460
840	5.1	3400

Determine the discharge in the creek.

SOLUTION:

The time versus salt content graph plotted below has an area of

$$A = 334,560 \text{ mg} \frac{\text{sec}}{\text{liter}}$$

By Equation 10.21,

$$Q = \frac{E}{A} = \frac{\text{mg of salt input}}{\text{mg(sec/liter)}}$$

$$= \frac{4.45 \times 10^5}{3.34 \times 10^5}$$

$$= 1.332 \frac{\text{liters}}{\text{sec}}$$

$$Q = 0.0471 \text{ cfs}$$

A simple but approximate information on the velocity in an irrigation or drainage canal may be obtained by *surface floats*. For streams of medium size sometimes 15 to 25 floats are used simultaneously. These are equally distributed along the starter channel cross section. At a distance away from the starter section an arrival section is located where the time of arrival of the floats are recorded. The distance between the two sections should provide a travel time of several minutes, unless the velocity of the stream is rapid and the channel is meandering. Since the float velocity is representative of the surface velocity of the stream, the measured velocities should be reduced by a factor ranging from 0.75 to 0.9, the larger value being applied to deeper and faster (more than 2.0 m/s) streams. Somewhat more precise results may be obtained by *adjustable submerging floats* that may be made by thin aluminum tubes closed at both ends and filled with weight at one end to the degree that they float. Such floats with lengths so selected that they submerge to a depth of about 10 to 15 in. above the bottom and stick out over the surface by about 2 to 4 in. The observed velocity of a submerging float, v_f, allows the determination of the average stream velocity, v_{ave}, along its path by the experimental formula

$$v_{ave} = v_f \left(0.9 - 0.116 \sqrt{1 - \frac{h_f}{y}} \right) \quad (10.23)$$

where h_f is the length of the float and y is the depth of the stream.

Float type measurements are sometimes used for large rivers also. This procedure may give valuable information on stream line patterns at a cost that is considerably less than procedures using current meters. For large dam projects in Europe and India floating torches were used at night and their paths were recorded by continuous photographing from several nearby hills. The results were reduced by using techniques known from descriptive geometry.

if they are of simple hydraulic design and if the water passes through them in critical condition. At *control structures*, weirs, gates, and the like, the discharge rating curve can be developed by a few independent discharge measurements in the stream, using the methods described in the previous sections.

DISCHARGE MEASUREMENT IN HYDRAULIC STRUCTURES

Hydraulic structures are ideal locations for discharge measurements, particularly

EXAMPLE 10.9

The discharge below a small dam was determined by repeated measurements using a propeller meter to be 6.5 m³/s on one occasion and 1.5 m³/s on another (Figure E10.9). The width of the weir was

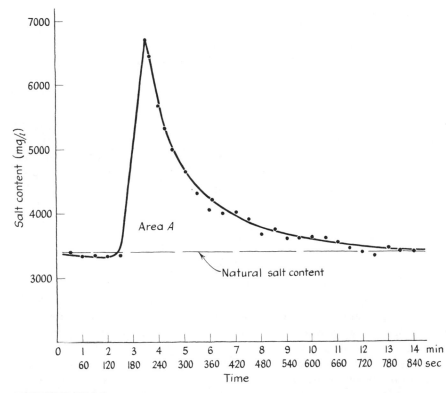

FIGURE E10.9

6 m on the first occasion, the depth of the water nappe over the weir was 57 cm; the second time the width was 21.5 cm. Design a rating curve for the weir.

SOLUTION:

From $Q = MbH^{3/2}$ and by Equation 9.10,

$$M_1 = \frac{Q_1}{bH_1^{3/2}}$$

$$= \frac{6.5}{6 \times (0.57)^{3/2}} = 2.52$$

$$M_2 = \frac{1.5}{6(0.215)^{3/2}} = 2.51$$

$$M_{ave} = \tfrac{1}{2}(2.52 + 2.51) = 2.515$$

then

$$Q = 2.515 \times 6H^{3/2}$$

$$Q = 15.09 H^{3/2}$$

Plotting this equation will provide the rating curve desired.

In installations where the amount of passing water must be known, as in irrigation, drainage, or sewage networks, permanent measuring structures are often built. Most often these structures are of standard configuration. Standardization makes their installation cost considerably less. Furthermore, the need for extensive calibration is eliminated if the structure is of well-known design.

One of the most successful structures developed for the measurement of small open channel flows is the *Parshall flume*, shown in Figure 10.22. Although more expensive than weirs the Parshall flume has found considerable use in irrigation, water supply, and waste water treatment plants because its inherent low permanent head loss and its self-cleaning capacity. Its main feature is its rapidly converging and dropping throat section in which critical flow conditions develop unless a hydraulic jump is formed by high downstream water level and the jump backs up into the throat section. For free flow in the throat the discharge in the flume can be computed by measuring the upstream water level in the stilling basin located on the outside of the converging side wall.

The approximate formula for the discharge in this case is

$$Q_0 = 4wH_a^{1.522w^{0.026}} \qquad (10.24)$$

where w is the throat width and H_a is the upstream water level. The nomograph shown in Figure 10.23 facilitates the solution of Equation 10.23. For larger discharges when the hydraulic jump moves upstream and completely submerges the throat section, the critical flow condition is eliminated. For this case, expressing the submergence by the ratio H_b/H_a in which H_b is measured at the throat section, the solution of the discharge may be corrected by the correction factor given by the graph shown in Figure 10.24.

Many *other venturi type flumes* were developed for various applications, each claiming certain advantages over the Parshall flume. For example, the Palmer-Bowlus flumes widely used in California and elsewhere for measuring sewage flows through manholes have a flat bottom; hence they are easier to place in existing installations. In some irrigation systems venturi flumes different from the Parshall flume are used with configurations that are cheaper to construct. The

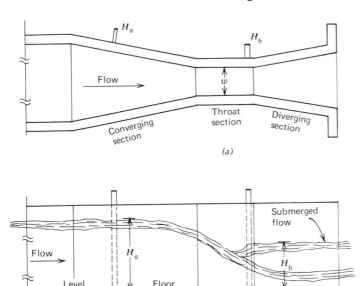

FIGURE 10.22 Parshall flume configuration. (*a*) Plan and (*b*) section. (Courtesy of Public Works Magazine.)

Balloffet measuring throat, consisting of a rectangular entrance chamber followed by a narrow rectangular throat section, is one such example that is in wide use in Argentina. Trapezoidal-shaped venturi flumes of various standard designs are used in Europe and elsewhere.

Regardless of their design, the well-designed discharge measuring structures' general requirements are that their discharge rating curve be independent of the channel slope and of downstream changes, such that their discharge be determined by a single depth measurement; that the changes in upstream and downstream channel configurations do not influence their discharge characteristics; that their size and configuration allow the designer to use established theoretical principles for preliminary design; that their discharge could be calibrated by only a few measurements in the field; and that their structure remain unchanged for a long period of time, hence their rating curve is reasonably permanent.

Measuring structures are generally installed on small streams. Since changes in flow in small streams are usually much more rapid than in large rivers, the occasional (daily or twice daily) reading of a graduated stage may introduce large errors in the continuous discharge data. This consideration suggests the necessity of installing *continuous recording devices* to measure the stage. Float type continuous recorders are available commercially. These recorders are either of the mechanical, clock motor driver type, like the *Stevens Total Flow Meter*, or electrical strip chart recorders. The Stevens flowmeter is equipped with interchangeable gears and cams to convert them to different flow ranges and include a paper carrying roller on which a pen marks the instantaneous position of the float. The roller is connected to a counter that integrates the "run" of the roller, hence, provides a reading that relates directly to the totalized volume of the flow.

Two types of floats are found in the

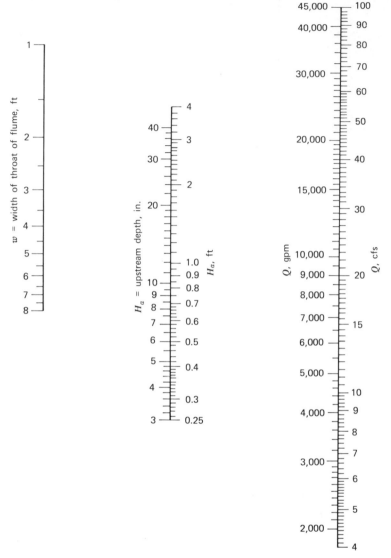

FIGURE 10.23   Discharge monograph for Parshall flumes. (Courtesy of Public Works
Magazine.)

field. One is the *drum float* that is used to measure elevations in still water. The other is the *scow float* that is a stream-lined float held by a rotating arm slanted in the direction of the stream. The scow type float allows stage measurements in flowing water. The arrangement of drum type and scow type floats is shown in Figure 10.25.

Electronic readout of the position of the water level is provided by the *NB Elec-* *tronic Manhole Meter*, which consists of a scow type arm equipped with a potentiometer type position indicating transducer at its pivot point. Since the resistance of the potentiometer is directly related to the position of the arm, this resistance value is recorded by a strip chart recorder. The *University of Akron scow float* is a similar device except that it includes a capacitance type network that allows the short term averaging and stor-

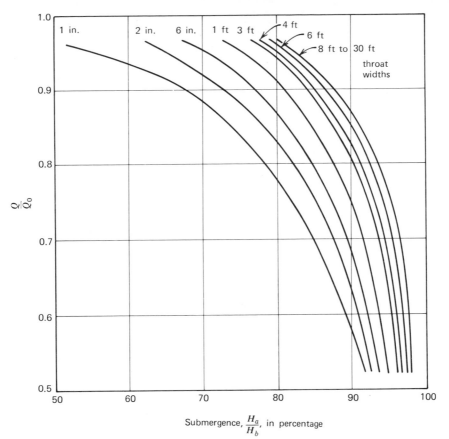

FIGURE 10.24   Correction factor for submerged Parshall flume's discharge determination. (Courtesy of Public Works Magazine.)

age of voltage fluctuation. This system is capable of measuring average water levels over several minute intervals in highly fluctuating waves found in energy dissipators. The electronic storage feature allows the use of a periodically (5 to 10 minutes) printing recorder, like the Esterline-Angus digital recorder.

An entirely different water level measuring system is utilized in the Swiss-made *Rittmeyer Hydrographic Water Level Recorder*. It is a pneumatic unit with a compressed air source supplying a slight air stream into the water measured through a control valve and a sensing tube. The air pressure in the sensing tube is proportional to the water level above the end of the tube. This pressure is recorded by a chart recorder or transmitted electrically to a distant control station.

REDUCTION AND PRESENTATION OF EXPERIMENTAL DATA

Hydraulics is an experimental science. Its development over the past two centuries was marked by the assemblage of an immense pile of experimental data from both field and laboratory measurements. Hydraulics is also a complex subject. Most phenomena involving the movement of water are known to be dependent on many variables: geometric characteristics, fluid properties, and flow characteristics. Even some of the most commonplace hydraulics problems depend on perhaps a dozen variables. Early books on hydraulics were replete with endless pages listing "raw" experimental data. Design consisted of selecting

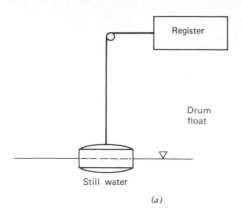

Register

Drum
float

Still water

(a)

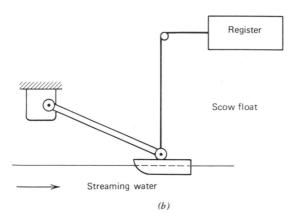

Register

Scow float

Streaming water

(b)

FIGURE 10.25   Depth measuring floats. (a) Drum float and (b) scow float.

parameters from tables showing data of previous projects "that worked." Much of this "raw" experimental data was incomplete, ambiguous, and incorrectly taken. Attempts to represent previous experiments by plotting them on charts and expressing the plots by empirical formulas increased the chances for errors by not stating the limits within which data was available, leading the users to incorrect extrapolations. To overcome manipulative difficulties inherent in handling large sets of experimental data, techniques of dimensional analysis and the formation of dimensionless groups of variables was introduced at the end of the nineteenth century. Dimensional analysis helps the experimenter in the systematic study of his experimental data. Dimensionless

grouping reduces the number of variables that has to be processed.

In physical systems, including hydraulic ones, certain quantities such as mass or force, length, time, and temperature are considered *fundamental quantities* that cannot be expressed in simple terms. These fundamental quantities are said to have dimensions. We measure these dimensions in various units; for instance, length may be measured in meters, light years, feet, or fathoms. All other physical quantities are expressed in terms of the fundamental quantities: velocity is expressed as length per time and so on. These are the *derived quantities*.

Since force and mass are related to each other by Newton's law of motion—force equals mass times acceleration—only four of the five fundamental quantities are independent from each other. To describe motion we either use the force, length, and time system or use the mass, length, and time system. In either case, the mass or force becomes a derived quantity. Either set may be used for dimensional analysis of physical systems, but they must be used exclusively. For both sets of fundamental quantities, derived dimensions of most common hydraulic parameters are listed in Table 10-2.

In order to best organize a set of "raw" experimental data or to plan an experimental study the first step is to list all physical variables that pertain to the problem. This requires from the analyst that he consider all fundamental theoretical concepts that play a part in the physical process studied. For example, hydraulic energy must be looked upon as foot-pounds of energy per pound of fluid, instead of its common representation as "feet of head." Terms not obviously playing a part in the problem must be listed if they inherently influence the physical processes considered. For example, the gravitational acceleration is essential in the analysis of weir flow, even though its value remains a constant, and as such the

Table 10-2 Dimensions of Hydraulic Quantities

	FLT	MLT
Geometrical characteristics		
Length	L	L
Area	L^2	L^2
Volume	L^3	L^3
Fluid Properties[a]		
Mass	FT^2/L	M
Density (ρ)	FT^2/L^4	M/L^3
Specific weight (γ)	F/L^3	M/L^2T^2
Kinematic viscosity (ν)	L^2/T	L^2/T
Absolute viscosity (μ)	FT/L^2	M/LT
Elastic modulus (K)	F/L^2	M/LT^2
Surface tension (σ)	F/L	M/T^2
Flow Characteristics		
Velocity (v)	L/T	L/T
Angular velocity (ω)	$1/T$	$1/T$
Acceleration (α)	L/T^2	L/T^2
Pressure (Δp)	F/L^2	M/LT^2
Force (drag, lift, shear)	F	ML/T^2
Shear stress (τ)	F/L^2	M/LT^2
Pressure gradient ($\Delta p/L$)	F/L^3	
Discharge	L^3/T	L^3/T
Mass flow rate	FT/L	M/T
Work or energy	FL	ML^2/T^2
Work or energy per unit weight	L	L
Torque and moment	FL	ML^2/T^2
Work or energy per unit mass	L^2/T^2	L^2/T^2

[a] Density, viscosity, elastic modulus, and surface tension depend upon temperature; therefore temperature will not be considered a property in the sense used here.

careless experimenter may tend to neglect it.

Glancing over the formulas and charts in the previous pages of this book, we notice that the advantages of dimensionless representation of data become obvious. Dimensionless groups of variables are independent of the units of measurement; for example, the Reynolds number is the same whether English or metric units are used to express it. (Needless to say that dimensional homogeneity is a fundamental requirement in all computations, since all variables have to be expressed in a consistent system of measures.)

As an example of *dimensional analysis* let us consider the case of flow in pipes. The pressure drop in a unit length of pipe, $\Delta p / L$, is dependent on the pipe diameter, D, the pipe roughness e, the average flow velocity v, the density of the fluid ρ, and its viscosity μ. Hence, we may write

$$\frac{\Delta p}{L} = f(D, e, v, \rho, \mu) \qquad (10.25)$$

Writing the dimensions in the force, length, and time system we get, by the aid of Table 10-2,

$$\frac{F}{L} = f\left[L, L, \left(\frac{L}{T}\right), \left(\frac{FT^2}{L^4}\right), \left(\frac{FT}{L^2}\right)\right] \qquad (10.26)$$

We observe that the formula involves all three of the fundamental quantities; hence, we write $m = 3$. Since there are six derived quantities in our problem, we write $n = 6$.

The fundamental theorem of dimensional analysis states that the minimum number of dimensionless units that can be composed from these variables equals $(n - m)$, which is in our case three. By tradition these dimensionless units are called Π (pi) terms. One systematic method to form these Π terms is called Buckingham's Π theorem.

The procedure of forming meaningful dimensionless variables consists of the following steps.

First, a set of m variables selected from

the ones in Equation 10.25 such that

(a) Among the three variables all of the fundamental quantities are involved
(b) One is a geometric characteristic, another is a fluid property, and the third is a flow characteristic
(c) They do not include the dependent variable (pressure loss per unit length in this case).

Next, the variables to be included in the $(n - m)$ dimensionless terms are collected such that each includes all previously picked variables plus one of the remaining ones. In our example we may select D, and v as repeating variables; hence the dimensionless terms will be composed of the following:

Π_1 will include D, ρ, v and $\dfrac{\Delta p}{L}$

Π_2 will include D, ρ, v and e

Π_3 will include D, ρ, v and μ

Now we may substitute their dimensions in place of their actual symbols from Equation 10.26 and write

$$\Pi_1 = (L)^x \left(\frac{FT}{L^4}\right)^y \left(\frac{L}{T}\right)^z \frac{F}{L^3} \qquad (10.27)$$

and so on for the other two Π terms.

In order to make Π_1 dimensionless the exponents x, y, and z should be selected so that Equation 10.27 is equal to one. As the zeroth power of any number is one, Equation 10.27 may be written as

$$\Pi_1 = F^0 L^0 T^0 = 1 \qquad (10.28)$$

As F, L, and T are fundamental quantities, Equations 10.27 and 10.28 may be split up in the form of

$$F^y F = F^0$$
$$L^x L^{-4y} L^z L^{-3} = L^0 \qquad (10.29)$$
$$T^{2y} T^{-z} = T^0$$

From here we may write for the exponents to satisfy the above equation that

$$y + 1 = 0 \qquad (10.30)$$
$$x - 4y + z - 3 = 0 \qquad (10.31)$$
$$2y - z = 0 \qquad (10.32)$$

From Equation 10.30 we conclude that $y = -1$. Substituting this into Equation 10.32, we find that $z = -2$ and solving Equation 10.31 we finally get $x = 1$. Introducing these results into Equation 10.27,

$$\Pi_1 = \frac{\Delta p}{L} \frac{D}{\rho v^2} \qquad (10.33)$$

which, in view of Equation 4.3 in which $\Delta p = \gamma \cdot h_L$, equals

$$\Pi_1 = \frac{f}{2} \qquad (10.34)$$

one-half of the friction factor.

By a similar derivation

$$\Pi_2 = \frac{e}{D} \qquad (10.35)$$

the relative roughness, Equation 4.6 and

$$\Pi_3 = \frac{\mu}{vD\rho} \qquad (10.36)$$

the reciprocal of the Reynolds number, Equation 4.4. Since the reciprocal, square, square root, or any function of a dimensionless number is also a dimensionless number we can conclude that Equation 10.25 is equal to

$$\frac{\Delta p}{L} = f[\epsilon, \mathbf{R}, f] \qquad (10.37)$$

or, in other words, the head loss in a pipe can be expressed in a dimensionless manner as the friction factor, the Reynolds number, and the relative roughness, as indeed it is by Moody's dimensionless diagram (Figure 4.2) and the Darcy-Weisbach equation (Equation 4.3).

In practice it is often unnecessary to carry out a detailed dimensional analysis to be able to present experimental results in a proper manner. For example, when the Chézy coefficient is sought in Equation 8.16 the proper dimensions of C can be directly determined by substituting the dimensions in place of the variables. This gives

$$\frac{L}{T} = [C]\sqrt{L} \qquad (10.38)$$

from which the dimensions of C will be

$$[C] = \frac{L^{0.5}}{T} = \sqrt{\text{acceleration}} \qquad (10.39)$$

Since this is the square root of the dimension of acceleration it suggests that by including the square root of the gravitational acceleration into C, we may represent our results by a new coefficient k such that

$$k = C\sqrt{g} \qquad (10.40)$$

where k is now a dimensionless number.

It must be observed that dimensional analysis never solves a hydraulic problem; the functional relationship between the variables in, for example, Equation 10.37 should be determined by experimental means. However, after dimensionless grouping of our variables, we reduce their number significantly, making them easier to handle and eliminating our dependence on the units of measurement (English or metric).

The functional relationship of measured variables may be obtained by graphical analysis of the results. The first step of graphical analysis is the plotting of data on a normal, arithmetic plotting paper. If the resultant shows most experimental points in a narrow band that may be approximated by a straight line, as shown in Figure 10.26, a linear empirical formula may be composed in the form of

$$y = a + bx \qquad (10.41)$$

where y and x are the variables plotted on the vertical and horizontal coordinate lines, a is the value of y for which x equals zero, that is, the intercept of the line with the vertical coordinate, and b is the slope of the line as shown in Figure 10.26. If there are more than three variables expressed by the graph then a and b may be defined for each of the plotted variables, with the final result giving a family of lines.

If the plot cannot be approximated by a straight line, the function is probably of exponential nature. It means that the equation may take the form of

$$y = ax^b \qquad (10.42)$$

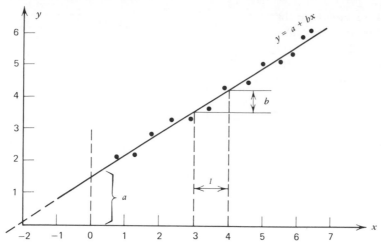

FIGURE 10.26   Plotting of data on normal graph paper.

with a and b being constants. Taking the logarithm of both sides of Equation 10.42 results in

$$\log y = \log a + b \log x \qquad (10.43)$$

which shows that on logarithmic paper the equation may plot as a straight line. The development of Equation 10.42 may proceed by first replotting the data on a log-log paper with a suitable number of cycles to encompass all experimental values. Then an approximating line may be drawn across the band of points. The value of $\log a$ may then be determined by reading the value on the y axis corresponding to the intercept of the line at the $\log x = 1$ position as shown in Figure 10.27, since $x = 0$ when $\log x = 1$. The value of b represents the slope of the line. For this two points x_1 and x_2 are selected on the line and their corresponding y_1 and y_2 values are read off the chart. The slope b is then expressed as

$$b = \frac{\log y_2 - \log y_1}{\log x_2 - \log x_1} \qquad (10.44)$$

If the graph shows a small curvature when plotted on a log-log paper the equation representing this graph may be expressed by introducing a small correction factor c such that

$$y = ax^b + c \qquad (10.45)$$

To find c first care should be taken to select x_1 and x_2 for Equation 10.44 near the extremities of the graph. Then a third point x_3 is selected as the geometric mean of x_1 and x_2, that is, $x_3 = \sqrt{x_1 \cdot x_2}$. The value of c is then determined by

$$c = \frac{y_1 y_2 - y_3^2}{y_1 + y_2 - 2y_3} \qquad (10.46)$$

where y_3 corresponds to x_3 on the curved line.

Some experimental results may be expressed in the form of

$$y = e^{a+bx} \qquad (10.47)$$

where e is the base of the Napierian (natural) logarithm, $e = 2.718$. If we know that Equation 10.47 may be written as

$$\log_e y = a + bx$$

and that

$$\log_{10} y = 0.434 \log_e y \qquad (10.48)$$

where $\log_{10}$ is the logarithm of the base of ten, Equation 10.47 may be dealt with more conveniently in the form of

$$y = 10^{a+bx}$$

where a and b are constants. The equation can be written as

$$\log_{10} y = a + bx \qquad (10.49)$$

for which semilogarithmic plotting papers are available.

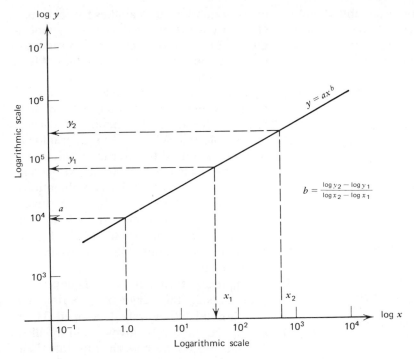

FIGURE 10.27 Plotting on logarithmic paper. $b = \log y_2 - \log y_1 / \log x_2 - \log x_1$.

If the experimental data plotted on a semilogarithmic paper results in a narrow band that can be approximated as a straight line as shown in Figure 10.28, the constant a can be found as the value of y for $x = 0$. The constant b is, again, the

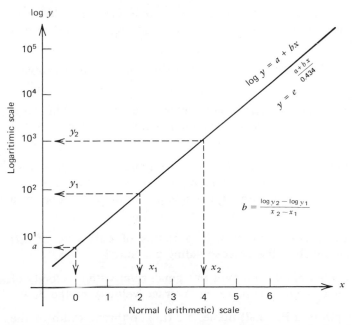

FIGURE 10.28 Plotting on semilogarithmic paper. $b = \log y_2 - \log y_1 / x_2 - x_1$.

slope of the line. For this suitable x_1 and x_2 values should be selected and the corresponding log y_1 and log y_2 values are read off the line. The slope b then will be

$$b = \frac{\log y_2 - \log y_1}{x_2 - x_1} \qquad (10.50)$$

Again, if the plot on the semilogarithmic graph does show a small curvature, the relationship may be linearized by introducing a correction factor c such that Equation 10.49 takes the form

$$\log (y - c) = a + bx \qquad (10.51)$$

The value of c can be determined by first selecting the x_1 and x_2 values for Equation 10.50 near the extremities of the plotted line and then finding x_3 such that

$$x_3 = \frac{x_1 + x_2}{2} \qquad (10.52)$$

for which y_3 may be determined from the graph. Equation 10.45 applies for this case also. In case of more than two variables the x and y terms may be formed as combinations of hydraulic variables. Also, for example, when three variables are present, one may be held constant at various values and the other two are plotted. By this method a family of curves may be plotted. When these semigraphical methods are used to determine functional relationships between hydraulic variables, it is advisable to try various combinations of x and y terms. In this sometimes tedious process one may attempt to plot variants of Π terms, such as $\Pi_1{}^{\Pi_2}$, Π_1/Π_3, tan Π, and so on, until the best possible fit is obtained.

Empirical formulas should never be used outside the range of experimental data for which they were developed. Their extrapolation rarely gives valid results. For this reason it is advisable to state their range of validity.

PROBLEMS

10.1 A circular orifice in an 8 in. pipe has a diameter of 4 in. Determine the velocity in the pipe if the discharge coefficient is 0.65 and the pressure differential measured is 18 psi.

10.2 Determine the K constant in Equation 10.7 for the orifice described in Problem 10.1.

10.3 A 4 ft long vertical differential mercury manometer registers $z = 12$ in. How much is the pressure measured in feet of water if the manometer is connected to the two pressure taps of an orifice meter?

10.4 A Pitot tube is inserted into an 8 in. diameter pipe. Assuming that at the center where the measurement is made the velocity is 25% larger than the average velocity, how high would the water rise in a vertical standpipe connected to the Pitot tube for a 1.5 cfs discharge?

10.5 A 6 in. diameter 90 degree elbow meter with a radius of curvature of 3 ft measured 4 cfs discharge. How much is the corresponding pressure?

10.6 A float submerging to a depth of 20 cm moves in a small river with a velocity of 80 cm/s. The depth of the river is 1.5 m. Estimate the average velocity of the flow.

10.7 The measured water depth in a Parshall flume is 2 ft. The throat width of the flume is 1 ft. How much is the discharge?

10.8   Perform a dimensional analysis on the hydraulics of sluice gates.

10.9   Develop a group of dimensionless numbers representing all variables related to a horizontal collector well.

10.10   Find an empirical equation for the graph shown in Figure 10.21.

APPENDIX

METRIC USAGE

The International System of Units (S.I.) is the metric system with precise standard definitions. These are agreed upon by the General Conference of the International Committee of Weights and Measures and recommended to the member countries for adoption. *Meter* is the fundamental unit of the metric system. Its precise definition is immaterial for us; for common usage it is enough to know that a meter is 1.09361 yard, or approximately 1.1 yard. The rest of the metric system can be simply derived from the meter.

Areas are measured by taking the square of the meter.

Volumes are measured by taking the cube of the meter. A commonly used metric measure of volume is the *liter*. It is a capacity of 1/10 of one meter cubed. It is slightly larger than a quart (1.05 qt).

Weights are defined by taking a unit volume and filling it with water. One cubic meter of water is one *metric ton*. One cubic centimeter of water is one *gram*. One cubic decimeter of water is one *kilogram*. One liter of water is one kilogram, since the volume of a liter is one cubic decimeter. The true standard of weight is the weight of a platinum-iridium bar kept in Paris.

Force in the S.I. System is expressed by the *newton*. By Newton's law mass times acceleration equals force. Kilogram is the metric unit of mass.

Prefixes like kilo-, deci-, etc. are widely used in the metric system. A prefix means multiplication or division of the basic unit by 10 or its multiples, 100, 1000, etc. Only a few of the many prefixes used in science are needed to be known by the common user; these terms fortunately are familiar to the English-speaking reader:

Kilo, as in kilogram, appears on the electric bill in the word of kilowatt, a metric unit already known to Americans meaning one thousand watts.

Centi, as in centimeter, is known from the term cent, 1/100 of a dollar.

Milli, as in milligram, appears in mil, known by mechanics as 1/1000 of an inch.

Deka, as in dekagram, is used in the word decade, meaning 10 years.

Deci, as in deciliter, is known from the word decimate, meaning to destroy every 10th one of something.

Prefixes used in scientific work are as

follows:

	Prefix	Symbol	Multiplies Basic Unit by
	deka	da	ten
	hecto	h	hundred
	kilo	k	thousand
Enlarging	mega	M	one million
Prefixes	giga	G	one thousand-million
	tera	T	one million-million

	Prefix	Symbol	Divides Basic Unit by
	deci	d	ten
	centi	c	hundred
	milli	m	thousand
	micro	μ	one million
Reducing	nano	n	one thousand million
Prefixes	pico	p	one million-million
	femto	f	one thousand-million-million
	atto	a	one million-million-million

Commonly used combinations of prefixes and corresponding metric units are shown below. These must be known by the users of the metric system.

early metric term, now abandoned. A hectare is an area 100 by 100 meters, or 10,000 square meters, which is about 2.5 acres.

Metric ton (t) is 1000 kilogram, almost equals the British (long) ton.

Abbreviations of metric units, shown above in parentheses after the name of each unit are to be considered symbols not unlike symbols used in chemistry or mathematics. Accordingly they are not to be succeeded by a period, unless they are at the end of a sentence. As precise scientific symbols they should be correctly adhered to. All of them are to be written in small case Roman letters (never in italics), unless they are derived from people's names, in which case they are in capital letters, (as kWh for kilowatthour). In compound units, for instance, km/h, kWh, and mm, they are to be written adjoiningly, without spaces, dots, or hyphens between them.

Compound, derived units, like areas and velocities: squaring or cubing may be signified by cu. for cubic, or sq. for square, but it is preferred that exponents be used, such as km^2, cm^2, etc. Since the exponent

Multiplier	Prefix Name	Corresponding Metric Units		
1000	kilo	kilometer (km)		kilogram (kg)
100	hecto	—	hectoliter (hl)	—
10	deka	—	—	dekagram
Unit Used =		*meter* (m)	*liter* (l)	*gram* (g)
1/10	deci	decimeter (dm)	deciliter (dl)	—
1/100	centi	centimeter (cm)	centiliter (cl)	—
1/1000	milli	millimeter (mm)	milliliter (ml)	milligram (mg)

Prefixes when used alone may have precise application. *Deci* may mean deciliter but never decimeter. *Centi* may mean centimeter but never centiliter. *Kilo* is kilogram, never kilometer. *Deka* may mean dekagram but never dekameter.

Hectare (ha) is a metric unit that is used as a land measure only. It is derived from the unit "are," a square dekameter, an

in such cases refer to the symbol as a whole, parenthesis, for example, in $(cm)^2$ is not called for.

Some common errors in metric abbreviations:

cc for cm^3

sec for s, although "sec" for second is common when used in connection with English notations.

gm or gr for g, gram.

KMH for kilometer per hour, used to imitate MPH. Correct term is km/h.

m/m for mm, millimeter.

The decimal nature of the metric system is one of its prime advantages. This advantage is negated if measurements are presented in a form in which the units are mixed. For example, it is incorrect to define $1\frac{1}{3}$ meter as "one meter, three decimeters, three centimeters, and three millimeters." The correct definition is "one-point-three three three meters", or, one hundred-thirty-three point three centimeters," that is 1.333 m, or 133.3 cm.

Selection of basic metric units to be used in expressing measured quantities depends on the magnitude of those quantities and their relationship to the metric units. The aim is to avoid using too many zeros. Distances on land are measured in kilometers, while the tailor uses centimeters and the machinist uses millimeters. The selection of the units inherently implies a degree of precision. A statement of 1.2 km defines the distances as less than 1.3 km but more than 1.1 km. The same distance defined as 1.200 km signifies that it is precise to the third digit behind the decimal point, that is, it is precise within a tolerance of one meter; therefore, it would be better expressed by writing it as 1200 m. Appendix 2 contains an extensive list of conversion factors for hydraulic terms giving the relationship between units of various systems of weights and measures. This list includes most metric conversion factors, but for those who desire to become familiar with the size of the metric (S.I.) units, the following approximate values may suffice.

1 millimeter	0.04 inch
1 meter	3.3 feet
1 meter	1.1 yard
1 kilometer	0.6 mile
1 square centimeter	0.16 square inch
1 square meter	11 square feet
1 square meter	1.2 square yards
1 hectare	2.5 acres
1 cubic centimeter	0.06 cubic inch
1 cubic meter	35 cubic feet
1 cubic meter	250 gallons
1 cubic meter	1.3 cubic yards
1 liter	1.05 quart
1 liter	3.63 cup
1 gram	0.035 ounce (avdp.)
1 kilogram	2.2 pounds

Celsius (centigrade) temperature = (Fahrenheit degree − 32) · 5/9

APPENDIX 2

VOLUME CONVERSION FACTORS

DISCHARGE CONVERSION FACTORS

RAINFALL AND RUNOFF CONVERSIONS

VOLUME CONVERSION FACTORS

	Cubic Inch	U.S. Gallon	Imperial Gallon	Cubic Foot	Cubic Yard	Cubic Meter	Acre-Foot	Second-Foot-Day
Cubic inch	1	0.00433	0.00361	5.79×10^{-4}	2.14×10^{-5}	1.64×10^{-5}	1.33×10^{-8}	6.70×10^{-9}
U.S. gallon	231	1	0.833	0.134	0.00495	0.00379	3.07×10^{-6}	1.55×10^{-6}
Imperial gallon	277	1.20	1	0.161	0.00595	0.00455	3.68×10^{-6}	1.86×10^{-6}
Cubic foot	1728	7.48	6.23	1	0.00370	0.0283	2.30×10^{-5}	1.16×10^{-5}
Cubic yard	46,656	202	168	27	1	0.765	6.20×10^{-4}	3.12×10^{-4}
Cubic meter	61,000	264	220	35.3	1.31	1	8.11×10^{-4}	4.09×10^{-4}
Acre-foot	7.53×10^{7}	3.26×10^{5}	2.71×10^{5}	43,560	1610	1230	1	0.504
Second-foot-day	1.49×10^{8}	6.46×10^{5}	5.38×10^{5}	86,400	3200	2450	1.98	1

DISCHARGE CONVERSION FACTORS

	Million Gallons/Day	Gallons/Minute	Cubic Foot/Second	Cubic Foot/Day	Acre-Foot/Day	Meter^3/Second	Meter^3/Hour	Meter^3/Day	Millimeter/Year/Kilometer^2	Inch/Year/Mile^2
Million gallons/day	1	694	1.55	134×10^3	307	43.8×10^{-3}	157.68	3784	1381	20.98
Gallons/minute	1.44×10^{-3}	1	2.227×10^{-3}	193	4.42×10^{-3}	6.31×10^{-5}	0.227	5.448	1.988	0.03
Cubic foot/second	0.646	449	1	86,400	1.98	0.0283	101.88	2445	892.4	13.56
Cubic foot/day	7.48×10^{-6}	5.19×10^{-3}	1.16×10^{-5}	1	2.3×10^{-5}	3.28×10^{-7}	1.18×10^{-3}	0.028	0.0102	1.55×10^{-4}
Acre-foot/day	0.326	226	0.504	43,560	1	0.0143	51.48	1235	450	6.84
Meter^3/second	22.8	15,800	35.3	3.05×10^6	70	1	3600	86,400	31,536	479
Meter^3/hour	6.34×10^{-3}	4.40	9.81×10^{-3}	847.5	0.0194	2.77×10^{-4}	1	24	8.76	0.133
Meter^3/day	2.64×10^{-4}	0.184	4.09×10^{-4}	35.7	8.1×10^{-4}	1.157×10^{-5}	41.6×10^{-3}	1	0.365	5.55×10^{-3}
Millimeter/year/kilometer^2	7.24×10^{-4}	0.503	1.12×10^{-3}	98	2.22×10^{-3}	3.17×10^{-5}	0.114	2.74	1	0.0152
Inch/year/mile^2	0.047	33.33	0.074	6451	0.146	2.09×10^{-3}	7.52	180	65.79	1

RAINFALL AND RUNOFF CONVERSIONS

	Millimeter/Year	Inch/Year	Liter/Second/Kilometer^2	Meter^3/Year/Kilometer^2	Foot^3/Second/Mile^2
Millimeter/year	1	0.0394	0.03169	1000	0.3468
Inch/year	25.4	1	0.805	25,400	8.81
Liter/second/kilometer^2	31.55	1.242	1	31,550	10.94
Meter^3/year/kilometer^2	0.001	0.392×10^{-4}	0.316×10^{-4}	1	3.47×10^{-4}
Foot^3/second/mile^2	2.88	0.113	0.0914	2880	1

APPENDIX 3

CHANNEL SECTION GEOMETRY

VALUES OF THE $N^{3/2}$ FUNCTION FOR N'S BETWEEN ZERO AND FIVE

VALUES OF THE $N^{5/2}$ FUNCTION

METRIC VALUES OF THE KINETIC ENERGY TERM, $v^2/2g$

THE $\sqrt{2gh}$ FUNCTION IN METRIC UNITS

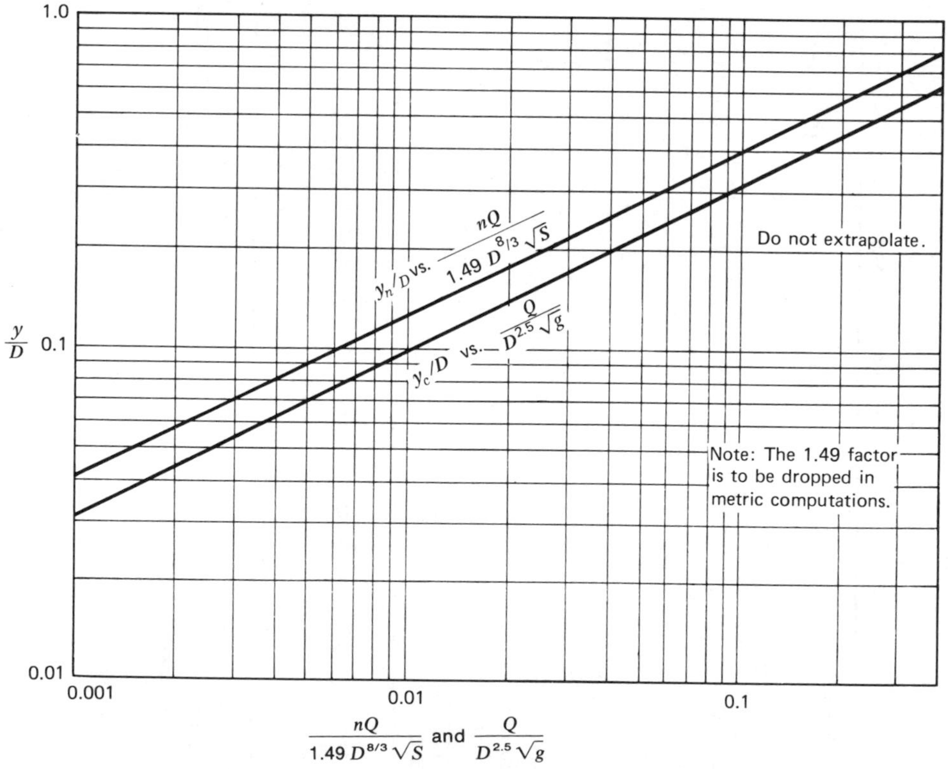

Critical and normal depths in partially full circular sections.

CHANNEL SECTION GEOMETRY

	Area A	Wetted Perimeter, P	Hydraulic Radius, $R = \dfrac{A}{P}$	Top Width, B	Depth, $\dfrac{A}{B}$	Critical Discharge,* Q_c
	$b \cdot y$	$b + 2y$	$\dfrac{by}{b+2y}$	b	y	$5.67 B y_c^{1.5}$
	$(b+zy)y$	$b + 2y\sqrt{1+z^2}$	$\dfrac{(b+zy)y}{b+2y\sqrt{1+z^2}}$	$b + 2zy$	$\dfrac{(b+zy)y}{b+2zy}$	$9.39[(b+zy_c)y_c]^{1.5}$
	zy^2	$2y\sqrt{1+z^2}$	$\dfrac{zy}{2\sqrt{1+z^2}}$	$2zy$	$0.5y$	$4zy_c^{2.5}$
	$0.125(\theta - \sin\theta)D^2$	$0.5\theta D$	$0.25\left(\dfrac{1-\sin\theta}{\theta}\right)D$	$2\sqrt{y(D-y)}$	$0.125\left(\dfrac{\theta-\sin\theta}{\sin\frac{1}{2}\theta}\right)D$	$\dfrac{0.251(\theta-\sin\theta)^{1.5}}{(\sin\frac{1}{2}\theta)^{0.5}}D^{2.5}$
Parabola	$\dfrac{2}{3}By$	$B + \dfrac{8y^2}{3B}$ **	$\dfrac{2B^2 y}{3B^2+8y^2}$ **	$\dfrac{3A}{2y}$	$\dfrac{2}{3}y$	$3.08 B y_c^{1.5}$

* For $g = 32.2$ ft/sec^2.
** For $y < B/4$.

VALUES OF THE $N^{3/2}$ FUNCTION FOR N'S BETWEEN ZERO AND FIVE*

N	0.00	0.01	0.02	0.03	0.04	0.05	0.06	0.07	0.08	0.09
0.0	0.0000	0.0010	0.0028	0.0052	0.0080	0.0112	0.0147	0.0185	0.0226	0.0270
0.1	.0316	.0365	.0416	.0469	.0524	.0381	.0640	.0701	.0764	.0828
0.2	.0894	.0962	.1032	.1103	.1176	.1250	.1326	.1403	.1482	.1562
0.3	.1643	.1726	.1810	.1896	.1983	.2071	.2160	.2251	.2342	.2436
0.4	.2530	.2625	.2722	.2820	.2919	.3019	.3120	.3222	.3326	.3430
0.5	.3536	.3642	.3750	.3858	.3968	.4079	.4191	.4303	.4417	.4532
0.6	.4648	.4764	.4882	.5000	.5120	.5240	.5362	.5484	.5607	.5732
0.7	.5857	.5983	.6109	.6237	.6366	.6495	.6626	.6757	.6889	.7023
0.8	.7155	.7290	.7425	.7562	.7699	.7837	.7975	.8115	.8255	.8396
0.9	.8538	.8681	.8824	.8969	.9114	.9259	.9406	.9553	.9702	.9850
1.0	1.000	1.015	1.030	1.045	1.061	1.076	1.091	1.107	1.122	1.138
1.1	1.154	1.169	1.185	1.201	1.217	1.233	1.249	1.266	1.282	1.298
1.2	1.315	1.331	1.348	1.364	1.381	1.398	1.414	1.431	1.448	1.465
1.3	1.482	1.499	1.517	1.534	1.551	1.569	1.586	1.604	1.621	1.639
1.4	1.657	1.674	1.692	1.710	1.728	1.746	1.764	1.782	1.800	1.819
1.5	1.837	1.856	1.874	1.893	1.911	1.930	1.948	1.967	1.986	2.005
1.6	2.024	2.043	2.062	2.081	2.100	2.119	2.139	2.158	2.178	2.197
1.7	2.217	2.236	2.256	2.275	2.295	2.315	2.335	2.355	2.375	2.395
1.8	2.415	2.435	2.455	2.476	2.496	2.516	2.537	2.557	2.578	2.598
1.9	2.619	2.640	2.660	2.681	2.702	2.723	2.744	2.765	2.786	2.807
2.0	2.828	2.850	2.871	2.892	2.914	2.935	2.957	2.978	3.000	3.021
2.1	3.043	3.065	3.087	3.109	3.131	3.153	3.175	3.197	3.219	3.241
2.2	3.263	3.285	3.308	3.330	3.353	3.375	3.398	3.420	3.443	3.465
2.3	3.488	3.511	3.534	3.557	3.580	3.602	3.626	3.649	3.672	3.695
2.4	3.718	3.741	3.765	3.788	3.811	3.835	3.858	3.882	3.906	3.929
2.5	3.953	3.977	4.000	4.024	4.048	4.072	4.096	4.120	4.144	4.168
2.6	4.192	4.217	4.241	4.265	4.289	4.314	4.338	4.363	4.387	4.412
2.7	4.437	4.461	4.486	4.511	4.536	4.560	4.585	4.610	4.635	4.660
2.8	4.685	4.710	4.736	4.761	4.786	4.811	4.837	4.862	4.888	4.913
2.9	4.939	4.964	4.990	5.015	5.041	5.067	5.093	5.118	5.144	5.170
3.0	5.196	5.222	5.248	5.274	5.300	5.327	5.353	5.379	5.405	5.432
3.1	5.458	5.485	5.511	5.538	5.564	5.591	5.617	5.644	5.671	5.698
3.2	5.724	5.751	5.778	5.805	5.832	5.859	5.886	5.913	5.940	5.968
3.3	5.995	6.022	6.049	6.077	6.104	6.132	6.159	6.186	6.214	6.242
3.4	6.269	6.297	6.325	6.352	6.380	6.408	6.436	6.464	6.492	6.520
3.5	6.548	6.576	6.604	6.632	6.660	6.689	6.717	6.745	6.774	6.802
3.6	6.831	6.859	6.888	6.916	6.945	6.973	7.002	7.031	7.059	7.088
3.7	7.117	7.146	7.175	7.204	7.233	7.262	7.291	7.320	7.349	7.378
3.8	7.408	7.437	7.466	7.495	7.525	7.554	7.584	7.613	7.643	7.672
3.9	7.702	7.732	7.761	7.791	7.821	7.850	7.880	7.910	7.940	7.970
4.0	8.000	8.030	8.060	8.090	8.120	8.150	8.181	8.211	8.241	8.272
4.1	8.302	8.332	8.363	8.393	8.424	8.454	8.485	8.515	8.546	8.577
4.2	8.607	8.638	8.669	8.700	8.731	8.762	8.793	8.824	8.855	8.886
4.3	8.917	8.948	8.979	9.010	9.041	9.073	9.104	9.135	9.167	9.198
4.4	9.230	9.261	9.293	9.324	9.356	9.387	9.419	9.451	9.482	9.514
4.5	9.546	9.578	9.610	9.642	9.674	9.705	9.737	9.770	9.802	9.834
4.6	9.866	9.898	9.930	9.963	9.995	10.03	10.06	10.09	10.12	10.16
4.7	10.19	10.22	10.25	10.29	10.32	10.35	10.39	10.42	10.45	10.48
4.8	10.52	10.55	10.58	10.62	10.65	10.68	10.71	10.75	10.78	10.81
4.9	10.85	10.88	10.91	10.95	10.98	11.01	11.05	11.08	11.11	11.15

* From the book of standards of the Research Institute of Water Resources, Budapest, Hungary, written by O. Starosolszky and L. Muszkalay, entitled *Mütárgyhidraulikai Zsebkönyv* (handbook of hydraulic structures, in Hungarian), Vols. 1 and 2, published by the "Technical Bookpublisher Co." Budapest, Hungary, 1961.

VALUES OF THE $N^{5/2}$ FUNCTION FOR N'S BETWEEN ZERO AND FIVE*

N	0.00	0.01	0.02	0.03	0.04	0.05	0.06	0.07	0.08	0.09
0.0	0.0000	0.0000	0.0001	0.0002	0.0003	0.0006	0.0009	0.0013	0.0018	0.0024
0.1	.0032	.0040	.0050	.0061	.0073	.0087	.0102	.0119	.0137	.0157
0.2	.0179	.0202	.0227	.0254	.0282	.0313	.0345	.0379	.0415	.0453
0.3	.0493	.0535	.0579	.0626	.0674	.0725	.0778	.0833	.0890	.0950
0.4	.1012	.1076	.1143	.1212	.1284	.1358	.1435	.1514	.1596	.1681
0.5	.1768	.1857	.1950	.2045	.2143	.2243	.2347	.2453	.2562	.2674
0.6	.2789	.2906	.3027	.3150	.3277	.3406	.3539	.3674	.3813	.3955
0.7	.4100	.4248	.4399	.4553	.4711	.4871	.5035	.5203	.5373	.5547
0.8	.5724	.5905	.6089	.6276	.6467	.6661	.6859	.7060	.7265	.7473
0.9	.7684	.7900	.8118	.8341	.8567	.8796	.9030	.9267	.9507	.9752
1.0	1.000	1.025	1.051	1.077	1.103	1.130	1.157	1.184	1.212	1.240
1.1	1.269	1.298	1.328	1.357	1.388	1.418	1.449	1.481	1.513	1.545
1.2	1.577	1.611	1.644	1.678	1.712	1.747	1.782	1.818	1.854	1.890
1.3	1.927	1.964	2.002	2.040	2.079	2.118	2.157	2.197	2.237	2.278
1.4	2.319	2.361	2.403	2.445	2.488	2.532	2.576	2.620	2.665	2.710
1.5	2.756	2.802	2.848	2.896	2.943	2.991	3.040	3.089	3.138	3.188
1.6	3.238	3.289	3.340	3.392	3.444	3.497	3.550	3.604	3.658	3.713
1.7	3.768	3.824	3.880	3.937	3.994	4.051	4.109	4.168	4.227	4.287
1.8	4.347	4.408	4.469	4.530	4.592	4.655	4.718	4.782	4.846	4.911
1.9	4.976	5.042	5.108	5.175	5.242	5.310	5.378	5.447	5.516	5.586
2.0	5.657	5.728	5.799	5.871	5.944	6.017	6.091	6.165	6.240	6.315
2.1	6.391	6.467	6.544	6.621	6.699	6.778	6.857	6.937	7.017	7.098
2.2	7.179	7.261	7.343	7.426	7.510	7.594	7.678	7.764	7.849	7.936
2.3	8.023	8.110	8.198	8.287	8.376	8.466	8.556	8.647	8.739	8.831
2.4	8.923	9.017	9.110	9.205	9.300	9.395	9.492	9.588	9.686	9.784
2.5	9.882	9.981	10.08	10.18	10.28	10.38	10.49	10.59	10.69	10.80
2.6	10.90	11.01	11.11	11.22	11.32	11.43	11.54	11.65	11.76	11.87
2.7	11.98	12.09	12.20	12.31	12.43	12.54	12.66	12.77	12.89	13.00
2.8	13.12	13.24	13.35	13.47	13.59	13.71	13.83	13.95	14.08	14.20
2.9	14.32	14.45	14.57	14.69	14.82	14.95	15.07	15.20	15.33	15.46
3.0	15.59	15.72	15.85	15.98	16.11	16.25	16.38	16.51	16.65	16.78
3.1	16.92	17.06	17.19	17.33	17.47	17.61	17.75	17.89	18.03	18.18
3.2	18.32	18.46	18.61	18.75	18.90	19.04	19.19	19.34	19.48	19.63
3.3	19.78	19.93	20.08	20.24	20.39	20.54	20.69	20.85	21.00	21.16
3.4	21.32	21.47	21.63	21.79	21.95	22.11	22.27	22.43	22.59	22.75
3.5	22.92	23.08	23.25	23.41	23.58	23.74	23.91	24.08	24.25	24.42
3.6	24.59	24.76	24.93	25.11	25.28	25.45	25.63	25.80	25.98	26.16
3.7	26.33	26.51	26.69	26.87	27.05	27.23	27.41	27.60	27.78	27.96
3.8	28.15	28.33	28.52	28.71	28.90	29.08	29.27	29.46	29.65	29.85
3.9	30.04	30.23	30.42	30.62	30.81	31.01	31.21	31.40	31.60	31.80
4.0	32.00	32.20	32.40	32.60	32.81	33.01	33.21	33.42	33.62	33.83
4.1	34.04	34.25	34.45	34.66	34.87	35.08	35.30	35.51	35.72	35.94
4.2	36.15	36.37	36.58	36.80	37.02	37.24	37.46	37.68	37.90	38.12
4.3	38.34	38.56	38.79	39.01	39.24	39.47	36.69	39.92	40.15	40.38
4.4	40.61	40.84	41.07	41.31	41.54	41.77	42.01	42.24	42.48	42.72
4.5	42.96	43.20	43.44	43.68	43.92	44.16	44.40	44.65	44.89	45.14
4.6	45.38	45.63	45.88	46.13	46.38	46.63	46.88	47.13	47.38	47.64
4.7	47.89	48.15	48.40	48.66	48.92	49.17	49.43	49.69	49.95	50.22
4.8	50.48	50.74	50.01	51.27	51.54	51.80	52.07	52.34	52.61	52.88
4.9	53.15	53.42	53.69	53.97	54.24	54.51	54.79	55.07	55.34	55.62

* From the book of standards of the Research Institute of Water Resources, Budapest, Hungary, written by O. Starosolszky and L. Muszkalay, entitled *Mütárgyhidraulikai Zsebkönyv* (handbook of hydraulic structures, in Hungarian), Vols. 1 and 2, published by the "Technical Bookpublisher Co." Budapest, Hungary, 1961.

METRIC VALUES OF THE KINETIC ENERGY TERM, $v^2/2g$*

v [m/s]	$\frac{v^2}{2g}$ [m]	v [m/s]	$\frac{v^2}{2g}$ [m]	v [m/s]	$\frac{v^2}{2g}$ [m]	v [m/s]	$\frac{v^2}{2g}$ [m]
0.01	0.0000051	0.41	0.00857	0.81	0.0336	1.21	0.0748
0.02	0.0000204	0.42	0.00897	0.82	0.0343	1.22	0.0760
0.03	0.0000459	0.43	0.00943	0.83	0.0352	1.23	0.0770
0.04	0.0000816	0.44	0.00989	0.84	0.0360	1.24	0.0783
0.05	0.0001275	0.45	0.0103	0.85	0.0368	1.25	0.0795
0.06	0.0001835	0.46	0.0108	0.86	0.0377	1.26	0.0808
0.07	0.000250	0.47	0.0113	0.87	0.0386	1.27	0.0821
0.08	0.000326	0.48	0.0118	0.88	0.0394	1.28	0.0835
0.09	0.000413	0.49	0.0122	0.89	0.0403	1.29	0.0850
0.10	0.000510	0.50	0.0128	0.90	0.0413	1.30	0.0863
0.11	0.000617	0.51	0.0133	0.91	0.0422	1.31	0.0875
0.12	0.000734	0.52	0.0138	0.92	0.0433	1.32	0.0889
0.13	0.000862	0.53	0.0143	0.93	0.0441	1.33	0.0903
0.14	0.00100	0.54	0.0149	0.94	0.0450	1.34	0.0915
0.15	0.00115	0.55	0.0154	0.95	0.0460	1.35	0.0929
0.16	0.00130	0.56	0.0160	0.96	0.0470	1.36	0.0943
0.17	0.00147	0.57	0.0166	0.97	0.0479	1.37	0.0955
0.18	0.00165	0.58	0.0172	0.98	0.0489	1.38	0.0970
0.19	0.00184	0.59	0.0177	0.99	0.0500	1.39	0.0985
0.20	0.00204	0.60	0.0184	1.00	0.0510	1.40	0.1000
0.21	0.00224	0.61	0.0190	1.01	0.0520	1.41	0.1015
0.22	0.00247	0.62	0.0196	1.02	0.0530	1.42	0.1030
0.23	0.00269	0.63	0.0202	1.03	0.0541	1.43	0.1040
0.24	0.00293	0.64	0.0208	1.04	0.0551	1.44	0.1055
0.25	0.00319	0.65	0.0216	1.05	0.0562	1.45	0.1070
0.26	0.00344	0.66	0.0222	1.06	0.0573	1.46	0.1085
0.27	0.00371	0.67	0.0229	1.07	0.0584	1.47	0.1100
0.28	0.00399	0.68	0.0236	1.08	0.0596	1.48	0.1115
0.29	0.00428	0.69	0.0242	1.09	0.0607	1.49	0.1132
0.30	0.00458	0.70	0.0250	1.10	0.0617	1.50	0.1148
0.31	0.00490	0.71	0.0257	1.11	0.0627	1.51	0.1162
0.32	0.00523	0.72	0.0264	1.12	0.0639	1.52	0.1180
0.33	0.00553	0.73	0.0271	1.13	0.0650	1.53	0.1192
0.34	0.00593	0.74	0.0280	1.14	0.0663	1.54	0.1210
0.35	0.00624	0.75	0.0287	1.15	0.0675	1.55	0.1225
0.36	0.00660	0.76	0.0295	1.16	0.0685	1.56	0.1240
0.37	0.00698	0.77	0.0303	1.17	0.0698	1.57	0.1255
0.38	0.00736	0.78	0.0311	1.18	0.0709	1.58	0.1270
0.39	0.00775	0.79	0.0319	1.19	0.0723	1.59	0.1290
0.40	0.00815	0.80	0.0326	1.20	0.0735	1.60	0.1305

* From the book of standards of the Research Institute of Water Resources, Budapest, Hungary, written by O. Starosolszky and L. Muszkalay, entitled *Mütárgyhidraulikai Zsebkönyv* (handbook of hydraulic structures, in Hungarian), Vols. 1 and 2, published by the "Technical Bookpublisher Co." Budapest, Hungary, 1961.

METRIC VALUES OF THE KINETIC ENERGY TERM, $v^2/2g$*
(*Cont'd*)

v [m/s]	$\frac{v^2}{2g}$ [m]	v [m/s]	$\frac{v^2}{2g}$ [m]	v [m/s]	$\frac{v^2}{2g}$ [m]	v [m/s]	$\frac{v^2}{2g}$ [m]
1.61	0.1320	2.30	0.269	4.55	1.06	6.80	2.36
1.62	0.1335	2.35	0.281	4.60	1.08	6.85	2.40
1.63	0.1355	2.40	0.293	4.65	1.10	6.90	2.42
1.64	0.1370	2.45	0.306	4.70	1.13	6.95	2.46
1.65	0.1385	2.50	0.319	4.75	1.15	7.00	2.50
1.66	0.1400	2.55	0.331	4.80	1.18	7.05	2.54
1.67	0.1420	2.60	0.344	4.85	1.20	7.10	2.57
1.68	0.1435	2.65	0.358	4.90	1.22	7.15	2.60
1.69	0.1453	2.70	0.371	4.95	1.25	7.20	2.64
1.70	0.1472	2.75	0.386	5.00	1.28	7.25	2.68
1.71	0.1490	2.80	0.399	5.05	1.30	7.30	2.71
1.72	0.1510	2.85	0.414	5.10	1.33	7.35	2.75
1.73	0.1525	2.90	0.428	5.15	1.35	7.40	2.80
1.74	0.1545	2.95	0.445	5.20	1.38	7.45	2.83
1.75	0.1560	3.00	0.458	5.25	1.40	7.50	2.87
1.76	0.1580	3.05	0.474	5.30	1.43	7.55	2.91
1.77	0.1595	3.10	0.490	5.35	1.45	7.60	2.95
1.78	0.1615	3.15	0.508	5.40	1.48	7.65	2.98
1.79	0.1635	3.20	0.523	5.45	1.52	7.70	3.03
1.80	0.1650	3.25	0.540	5.50	1.54	7.75	3.06
1.81	0.1670	3.30	0.553	5.55	1.57	7.80	3.11
1.82	0.1690	3.35	0.572	5.60	1.60	7.85	3.14
1.83	0.1705	3.40	0.593	5.65	1.63	7.90	3.19
1.84	0.1725	3.45	0.607	5.70	1.66	7.95	3.22
1.85	0.1742	3.50	0.624	5.75	1.69	8.00	3.26
1.86	0.1762	3.55	0.642	5.80	1.72	8.05	3.31
1.87	0.1785	3.60	0.660	5.85	1.76	8.10	3.36
1.88	0.1805	3.65	0.678	5.90	1.77	8.15	3.39
1.89	0.1825	3.70	0.698	5.95	1.80	8.20	3.43
1.90	0.1845	3.75	0.719	6.00	1.84	8.25	3.47
1.91	0.1862	3.80	0.736	6.05	1.86	8.30	3.52
1.92	0.1880	3.85	0.755	6.10	1.90	8.35	3.56
1.93	0.1900	3.90	0.775	6.15	1.93	8.40	3.60
1.94	0.1920	3.95	0.795	6.20	1.96	8.45	3.64
1.95	0.1937	4.00	0.815	6.25	1.99	8.50	3.68
1.96	0.1958	4.05	0.835	6.30	2.02	8.55	3.72
1.97	0.1980	4.10	0.857	6.35	2.06	8.60	3.77
1.98	0.2000	4.15	0.877	6.40	2.08	8.65	3.82
1.99	0.2022	4.20	0.897	6.45	2.12	8.70	3.86
2.00	0.2040	4.25	0.923	6.50	2.16	8.75	3.90
2.05	0.214	4.30	0.943	6.55	2.18	8.80	3.94
2.10	0.224	4.35	0.963	6.60	2.22	8.85	3.98
2.15	0.236	4.40	0.989	6.65	2.25	8.90	4.03
2.20	0.247	4.45	1.010	6.70	2.29	8.95	4.08
2.25	0.258	4.50	1.030	6.75	2.32	9.00	4.13

METRIC VALUES OF THE KINETIC ENERGY TERM, $v^2/2g$*
(*Cont'd*)

v [m/s]	$\frac{v^2}{2g}$ [m]	v [m/s]	$\frac{v^2}{2g}$ [m]	v [m/s]	$\frac{v^2}{2g}$ [m]	v [m/s]	$\frac{v^2}{2g}$ [m]
9.05	4.17	9.30	4.41	9.55	4.64	9.80	4.89
9.10	4.22	9.35	4.46	9.60	4.69	9.85	4.94
9.15	4.27	9.40	4.50	9.65	4.74	9.90	5.00
9.20	4.32	9.45	4.55	9.70	4.79	9.95	5.05
9.25	4.36	9.50	4.60	9.75	4.84	10.00	5.10

THE $\sqrt{2gh}$ FUNCTION IN METRIC UNITS*

h [m]	$\sqrt{2gh}$ [m/s]	h [m]	$\sqrt{2gh}$ [m/s]	h [m]	$\sqrt{2gh}$ [m/s]	h [m]	$\sqrt{2gh}$ [m/s]
0.01	0.443	0.41	2.83	0.81	3.99	1.21	4.87
0.02	0.627	0.42	2.87	0.82	4.02	1.22	4.89
0.03	0.768	0.43	2.90	0.83	4.04	1.23	4.91
0.04	0.887	0.44	2.93	0.84	4.06	1.24	4.93
0.05	0.990	0.45	2.97	0.85	4.08	1.25	4.95
0.06	1.085	0.46	3.00	0.86	4.10	1.26	4.97
0.07	1.170	0.47	3.04	0.87	4.13	1.27	4.99
0.08	1.250	0.48	3.07	0.88	4.15	1.28	5.01
0.09	1.33	0.49	3.10	0.89	4.18	1.29	5.03
0.10	1.40	0.50	3.13	0.90	4.20	1.30	5.05
0.11	1.47	0.51	3.16	0.91	4.22	1.31	5.07
0.12	1.54	0.52	3.19	0.92	4.25	1.32	5.09
0.13	1.60	0.53	3.23	0.93	4.27	1.33	5.11
0.14	1.66	0.54	3.26	0.94	4.30	1.34	5.13
0.15	1.72	0.55	3.28	0.95	4.32	1.35	5.15
0.16	1.77	0.56	3.31	0.96	4.34	1.36	5.17
0.17	1.83	0.57	3.34	0.97	4.36	1.37	5.19
0.18	1.88	0.58	3.37	0.98	4.39	1.38	5.21
0.19	1.93	0.59	3.40	0.99	4.41	1.39	5.23
0.20	1.98	0.60	3.43	1.00	4.43	1.40	5.24
0.21	2.03	0.61	3.46	1.01	4.45	1.41	5.26
0.22	2.08	0.62	3.49	1.02	4.47	1.42	5.28
0.23	2.13	0.63	3.51	1.03	4.50	1.43	5.30
0.24	2.17	0.64	3.54	1.04	4.52	1.44	5.31
0.25	2.21	0.65	3.57	1.05	4.54	1.45	5.33
0.26	2.26	0.66	3.60	1.06	4.56	1.46	5.35
0.27	2.30	0.67	3.62	1.07	4.59	1.47	5.37
0.28	2.34	0.68	3.65	1.08	4.61	1.48	5.38
0.29	2.39	0.69	3.68	1.09	4.63	1.49	5.40
0.30	2.43	0.70	3.70	1.10	4.65	1.50	5.42
0.31	2.47	0.71	3.73	1.11	4.67	1.51	5.44
0.32	2.50	0.72	3.76	1.12	4.69	1.52	5.46
0.33	2.54	0.73	3.78	1.13	4.71	1.53	5.48
0.34	2.58	0.74	3.81	1.14	4.73	1.54	5.50
0.35	2.62	0.75	3.84	1.15	4.75	1.55	5.52
0.36	2.66	0.76	3.87	1.16	4.77	1.56	5.53
0.37	2.69	0.77	3.89	1.17	4.79	1.57	5.55
0.38	2.73	0.78	3.92	1.18	4.81	1.58	5.57
0.39	2.76	0.79	3.95	1.19	4.83	1.59	5.59
0.40	2.80	0.80	3.97	1.20	4.85	1.60	5.60

* From the book of standards of the Research Institute of Water Resources, Budapest, Hungary, written by O. Starosolszky and L. Muszkalay, entitled *Mütárgyhidraulikai Zsebkönyv* (handbook of hydraulic structures, in Hungarian), Vols. 1 and 2, published by the "Technical Bookpublisher Co." Budapest, Hungary, 1961.

THE $\sqrt{2gh}$ FUNCTION IN METRIC UNITS* (*Cont'd*)

h [m]	$\sqrt{2gh}$ [m/s]	h [m]	$\sqrt{2gh}$ [m/s]	h [m]	$\sqrt{2gh}$ [m/s]	h [m]	$\sqrt{2gh}$ [m/s]
1.61	5.62	2.30	6.72	4.55	9.45	6.80	11.54
1.62	5.64	2.35	6.79	4.60	9.50	6.85	11.58
1.63	5.66	2.40	6.87	4.65	9.55	6.90	11.62
1.64	5.67	2.45	6.94	4.70	9.60	6.95	11.66
1.65	5.69	2.50	7.02	4.75	9.65	7.00	11.70
1.66	5.71	2.55	7.09	4.80	9.70	7.05	11.74
1.67	5.73	2.60	7.15	4.85	9.75	7.10	11.78
1.68	5.75	2.65	7.22	4.90	9.80	7.15	11.82
1.69	5.77	2.70	7.28	4.95	9.85	7.20	11.86
1.70	5.78	2.75	7.35	5.00	9.90	7.25	11.90
1.71	5.80	2.80	7.42	5.05	9.94	7.30	11.94
1.72	5.82	2.85	7.48	5.10	9.99	7.35	11.98
1.73	5.83	2.90	7.55	5.15	10.03	7.40	12.02
1.74	5.85	2.95	7.61	5.20	10.08	7.45	12.06
1.75	5.87	3.00	7.68	5.25	10.12	7.50	12.10
1.76	5.88	3.05	7.74	5.30	10.17	7.55	12.14
1.77	5.90	3.10	7.80	5.35	10.21	7.60	12.18
1.78	5.92	3.15	7.86	5.40	10.26	7.65	12.22
1.79	5.94	3.20	7.93	5.45	10.30	7.70	12.26
1.80	5.95	3.25	7.99	5.50	10.35	7.75	12.30
1.81	5.97	3.30	8.05	5.55	10.40	7.80	12.34
1.82	5.98	3.35	8.11	5.60	10.45	7.85	12.38
1.83	5.99	3.40	8.17	5.65	10.50	7.90	12.42
1.84	6.01	3.45	8.23	5.70	10.55	7.95	12.46
1.85	6.03	3.50	8.29	5.75	10.60	8.00	12.50
1.86	6.04	3.55	8.35	5.80	10.63	8.05	12.54
1.87	6.06	3.60	8.41	5.85	10.70	8.10	12.58
1.88	6.07	3.65	8.46	5.90	10.75	8.15	12.62
1.89	6.09	3.70	8.52	5.95	10.80	8.20	12.66
1.90	6.10	3.75	8.58	6.00	10.85	8.25	12.70
1.91	6.12	3.80	8.63	6.05	10.90	8.30	12.74
1.92	6.13	3.85	8.69	6.10	10.94	8.35	12.78
1.93	6.15	3.90	8.75	6.15	10.99	8.40	12.82
1.94	6.17	3.95	8.81	6.20	11.03	8.45	12.86
1.95	6.19	4.00	8.87	6.25	11.08	8.50	12.90
1.96	6.20	4.05	8.92	6.30	11.12	8.55	12.94
1.97	6.22	4.10	8.97	6.35	11.17	8.60	12.98
1.98	6.23	4.15	9.03	6.40	11.22	8.65	13.02
1.99	6.25	4.20	9.08	6.45	11.26	8.70	13.06
2.00	6.27	4.25	9.14	6.50	11.30	8.75	13.10
2.05	6.24	4.30	9.19	6.55	11.34	8.80	13.14
2.10	6.42	4.35	9.24	6.60	11.38	8.85	13.18
2.15	6.49	4.40	9.30	6.65	11.42	8.90	13.22
2.20	6.57	4.45	9.35	6.70	11.46	8.95	13.26
2.25	6.64	4.50	9.40	6.75	11.50	9.00	13.30

THE $\sqrt{2gh}$ FUNCTION IN METRIC UNITS* (*Cont'd*)

h [m]	$\sqrt{2gh}$ [m/s]	h [m]	$\sqrt{2gh}$ [m/s]	h [m]	$\sqrt{2gh}$ [m/s]	h [m]	$\sqrt{2gh}$ [m/s]
9.05	13.33	9.30	13.50	9.55	13.69	9.80	13.86
9.10	13.37	9.35	13.54	9.60	13.72	9.85	13.90
9.15	13.40	9.40	13.57	9.65	13.76	9.90	13.93
9.20	13.44	9.45	13.61	9.70	13.79	9.95	13.97
9.25	13.47	9.50	13.65	9.75	13.83	10.00	14.00

BIBLIOGRAPHY

Addison, H. *A Treatise on Applied Hydraulics*. (Fifth Ed.) London: Chapman & Hall Ltd., 1964.

Albertson, M. L., J. R. Barton, and D. B. Simons. *Fluid Mechanics for Engineers*. Englewood Cliffs, N.J.: Prentice-Hall, Inc., 1960.

American Concrete Pipe Association. *Concrete Pipe Design Manual*. (First Ed.) 1970.

American Society of Civil Engineers. *Hydraulic Models*, ASCE Manual of Engineering Practice No. 25. New York, 1963.

American Society of Civil Engineers. *Hydrology Handbook*, ASCE Manual of Engineering Practice No. 28. New York, 1963.

Bean, H. S., ed. *Fluid Meters*. ASME Report. (Sixth Ed.) New York: American Society of Mechanical Engineers, 1971.

Cedergren, H. R. *Seepage, Drainage and Flow Nets*. New York: John Wiley & Sons, 1967.

Chow, V. T. *Open Channel Hydraulics*. New York: McGraw-Hill Book Co., 1959.

Davis, C. V. *Handbook of Applied Hydraulics*. (Second Ed.) New York: McGraw-Hill Book Co., 1952.

C. T. B. Donkin. *Elementary Practical Hydraulics of Flow in Pipes.* New Jersey: Oxford University Press, 1959.

Fair, G. M., and J. C. Geyer. *Water Supply and Waste-Water Disposal.* New York: John Wiley & Sons, 1956.

Finch, V. C. *Pump Handbook.* California: National Press, 1948.

Fox, R. W., and A. T. McDonald. *Introduction to Fluid Mechanics.* New York: John Wiley & Sons, 1973.

Graf, W. H. *Hydraulics of Sediment Transport.* New York: McGraw-Hill Book Co., 1971.

Harr, M. E. *Groundwater and Seepage.* New York: McGraw-Hill Book Co., 1962.

Hicks, T. G., and T. W. Edwards. *Pump Application Engineering.* New York: McGraw-Hill Book Co., 1971.

Jaeger, Charles. *Engineering Fluid Mechanics.* Translated from the German by P. O. Wolf. London: Blackie & Son, Ltd., 1956.

Johnstone, D., and W. P. Cross. *Elements of Applied Hydrology.* New York: The Ronald Press Co., 1949.

Kaufmann, W. *Fluid Mechanics.* New York: McGraw-Hill Book Co., 1963.

King, H. W., and E. F. Brater. *Handbook of Hydraulics.* (Fifth Ed.) New York: McGraw-Hill Book Co., 1963.

Kinori, B. Z. *Manual of Surface Drainage Engineering*, Vol. 1. Amsterdam: Elsevier Publishing Co., 1970.

Lea, F. C. *Hydraulics for Engineers and Engineering Students.* (Fourth Ed.) London: Edward Arnold & Co., 1923.

Leliavsky, S. *River and Canal Hydraulics.* London: Chapman and Hall, Ltd., 1965.

Linsley, R. K., M. A. Kohler, and J. L. H. Paulhus. *Applied Hydrology.* New York: McGraw-Hill Book Co., 1949.

Linsley, R. K., and J. B. Franzini. *Water Resources Engineering.* (Second Ed.) New York: McGraw-Hill Book Co., 1964.

Luthin, J. N. *Drainage Engineering.* New York: John Wiley & Sons, Inc., 1966.

Medaugh, F. W. *Elementary Hydraulics.* New York: D. Van Nostrand Co., 1924.

Mosonyi, E. *Water Power Development*, Vols. 1 and 2. Publishing House of Hungarian Academy of Science, Budapest, 1963.

Morris, H. M. *Applied Hydraulics in Engineering.* New York: The Ronald Press Co., 1963.

O'Brien, M. P., and George H. Hickox. *Applied Fluid Mechanics.* New York: McGraw-Hill Book Co., 1937.

Owczarek, J. A. *Introduction to Fluid Mechanics.* Scranton, Pa.: International Textbook Co., 1968.

Parmakian, J. *Waterhammer Analysis.* Englewood Cliffs, N.J.: Prentice-Hall, Inc., 1955.

Portland Cement Association. *Handbook of Concrete Culvert Pipe Hydraulics.* 1964.

Powell, R. W. *An Elementary Text in Hydraulics and Fluid Mechanics.* New York: The Macmillan Co., 1951.

Powell, R. W. *Mechanics of Liquids.* New York: The Macmillan Co., 1940.

Raghunath, H. M. *Dimensional Analysis and Hydraulic Model Testing.* Asia Publishing House, 1967.

Rouse, H. *Engineering Hydraulics.* New York: John Wiley & Sons, 1953.

Russell, G. E. *Hydraulics.* (Fifth Ed.) New York: H. Holt and Co., 1942.

Sellin, R. H. J. *Flow in Channels.* London: Gordon and Breach Science Publishers, 1970.

Spink, L. K. *Principles and Practice of Flow Meter Engineering.* (Ninth Ed.) The Foxboro Co., 1967.

Streeter, V. L. *Fluid Mechanics.* (Fourth Ed.) New York: McGraw-Hill Book Co., 1966.

Streeter, V. L. *Handbook of Fluid Dynamics.* New York: McGraw-Hill Book Co., 1961.

U.S. Department of Agriculture. *National Engineering Handbook.* Soil Conservation Service: Washington, D.C., 1971.

Vennard, J. K. *Elementary Fluid Mechanics.* (Third Ed.) New York, London: John Wiley & Sons; Chapman and Hall, Ltd., 1954.

Verruijt, A. *Theory of Groundwater Flow.* New York: The Macmillan Co., 1970.

Woodward, S. M. and C. J. Posey. *Hydraulics of Steady Flow in Open Channels.* New York: John Wiley & Sons, 1941.

INDEX

305